Will I Be Replaced by AI?

How AI Is Changing Our Workforce

Copyright ©SparklingVista2024
all rights reserved
ISBN: 9798326583307

Contents

Introduction..5

Chapter 1. Understanding the AI Revolution
-Overview of AI and automation..9
-The Evolution of AI and Automation...................................10
-Challenges and Considerations...12

Chapter 2. The Current Landscape: AI in Today's Workforce
-Examples of AI and automation in various industries...................14
-Case studies of companies implementing AI............................19
-The state of the job market..24

Chapter 3. Automation in Action: Transforming Industries
-Manufacturing and robotics...29
-AI in healthcare...33
-The impact on finance and banking....................................38
-Agriculture and smart farming..43

Chapter 4. Skills for the Future: Education and Training
-Identifying future-proof skills......................................49
-The role of education systems..54
-Lifelong learning and reskilling.....................................59

Chapter 5. Opportunities and Challenges: The Dual Impact of AI
-Job displacement vs. job creation....................................64
-The gig economy and new employment models............................68
-Addressing inequality and access to opportunities....................73

Chapter 6. Policy and Regulation: Governing the AI Era
-Government roles and responsibilities..80
-Regulatory frameworks and ethical guidelines..............................85

Chapter 7. Remote Work and Flexibility: The New Norm?
-Rise of remote work due to technology...91
-Benefits and drawbacks of remote and flexible work....................96
-Future projections for workplace flexibility.......................................101

Chapter 8. Mental Health and Well-being: Navigating Change
-Impact of automation on mental health...106
-Strategies for maintaining well-being..110
-Organizational responsibility and support systems........................115

Chapter 9. Ethics and Responsibility: AI in the Workplace
-Ethical considerations and dilemmas..120
-Corporate social responsibility..124
-Ensuring fairness and transparency..129

Chapter 10. The Road Ahead: Preparing for an AI-Driven Future
-Predictions and trends for the next decade....................................134
-Strategies for individuals and organizations...................................139
-Building a resilient and adaptable workforce................................144

Conclusion: Embracing Change and Innovation
-Recap of key insights...150
-The human element in the age of AI...154
-Final thoughts and calls to action..159

INTRODUCTION

Understanding the AI Revolution

In a world where technology evolves at breakneck speed, one question looms large for many of us: "Will I be replaced by AI?" This book seeks to answer that question by exploring the future of work in the age of automation and artificial intelligence. As we stand on the cusp of a new era, it's essential to understand how we got here and where we might be headed.

The Origins of Artificial Intelligence

The concept of artificial intelligence, or AI, has fascinated humanity for centuries. Early ideas can be traced back to ancient myths and stories of artificial beings endowed with intelligence by their creators. However, the formal study of AI began in the mid-20th century, with the pioneering work of mathematicians and computer scientists like Alan Turing and John McCarthy.

In 1950, Alan Turing proposed the famous Turing Test, a method to determine whether a machine could exhibit intelligent behavior indistinguishable from that of a human. This marked one of the first serious considerations of machine intelligence. Six years later, John McCarthy coined the term "artificial intelligence" and organized the Dartmouth Conference, which is often considered the birth of AI as a field of study. Early AI research focused on problem-solving and symbolic methods, leading to the development of the first AI programs and algorithms.

The Evolution of AI

The journey from these early days to the present has been marked by alternating periods of optimism and setbacks, often referred to as "AI winters" and "AI springs." During the 1970s and 1980s, AI faced significant challenges due to limited computing power and overly ambitious expectations, resulting in reduced funding and interest. However, the field revived in the late 1990s and early 2000s, thanks to advances in computational power, data availability, and new algorithms.

One of the most significant breakthroughs came with the development of machine learning, a subset of AI that enables systems to learn and improve from experience without being explicitly programmed. This was further accelerated by the advent of deep learning, which utilizes neural networks to process large amounts of data in complex ways. Landmarks such as IBM's Deep Blue defeating chess grandmaster Garry Kasparov in 1997, and Google DeepMind's AlphaGo beating Go champion Lee Sedol in 2016, showcased AI's potential to tackle complex tasks once thought to be uniquely human.

AI in the Modern Era

Today, AI and automation are embedded in our daily lives, often in ways we may not even realize. From voice assistants like Siri and Alexa to recommendation systems on Netflix and Amazon, AI shapes our interactions and decisions. In the workplace, AI is transforming industries by improving efficiency, reducing costs, and enabling new forms of innovation.

However, this rapid advancement also raises critical questions and challenges. As AI continues to evolve, it has the potential to displace certain jobs while creating new opportunities in others. This dual impact underscores the need for a nuanced understanding of AI's role in the future of work.

Purpose of This Book

"Will I Be Replaced by AI?" aims to demystify the complex and often misunderstood world of artificial intelligence and automation. Through this book, we will explore how different industries are being transformed, the skills that will be essential for the future, the ethical and regulatory landscapes, and the impact on mental health and well-being. By examining both the opportunities and challenges presented by AI, we hope to provide a comprehensive guide to navigating the changing world of work.

Join us as we delve into the fascinating journey of AI, from its origins to its current state, and explore what lies ahead in the age of automation. Together, we can embrace the changes and innovations that AI brings, ensuring a future where humans and machines coexist and thrive.

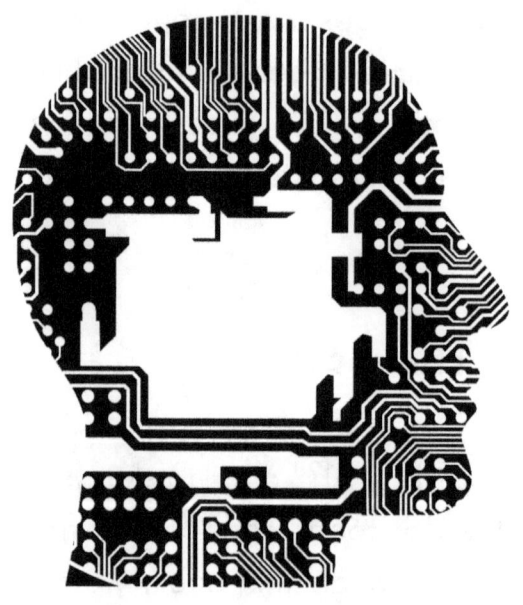

Chapter 1.
Understanding the AI Revolution

Overview of AI and automation

Artificial Intelligence (AI) and automation are two of the most transformative technologies shaping our present and future. Their influence spans across industries, altering how we work, live, and interact with the world. Understanding their scope, capabilities, and implications is crucial for navigating the modern technological landscape.

Defining AI and Automation

AI refers to the simulation of human intelligence in machines designed to think and learn like humans. It encompasses a variety of subfields, including machine learning, natural language processing, computer vision, and robotics. These technologies enable machines to perform tasks that typically require human intelligence, such as recognizing speech, identifying patterns, solving problems, and making decisions.

Automation, on the other hand, involves the use of technology to perform tasks with minimal human intervention. While it has been a part of industrial processes for decades, modern automation leverages AI to handle more complex and cognitive tasks. This integration allows for adaptive, flexible, and intelligent automation systems capable of improving efficiency and productivity across various domains.

The Evolution of AI and Automation

The journey of AI and automation began with simple mechanical devices and evolved through several significant milestones. Early automation involved machinery designed for repetitive tasks, such as assembly lines in manufacturing. The advent of computers in the mid-20th century brought a new wave of automation, enabling more sophisticated data processing and control systems.

The concept of AI emerged around the same time, with pioneers like Alan Turing and John McCarthy laying the groundwork. Initial AI systems were rule-based and limited in scope, relying on predefined algorithms and logic. However, advancements in machine learning, particularly deep learning, revolutionized AI by allowing systems to learn from data and improve over time.

Key Components of AI and Automation

1. Machine Learning (ML):
A subset of AI, ML involves training algorithms to learn from and make predictions based on data. It includes supervised learning (learning from labeled data), unsupervised learning (identifying patterns in unlabeled data), and reinforcement learning (learning through trial and error).

2. Natural Language Processing (NLP):
This area of AI focuses on enabling machines to

understand, interpret, and respond to human language. NLP powers applications such as chatbots, virtual assistants, and language translation services.

3. Computer Vision:
This field involves teaching machines to interpret and understand visual information from the world, such as images and videos. Applications include facial recognition, autonomous vehicles, and medical imaging.

4. Robotics:
Robotics combines AI with mechanical engineering to create machines capable of performing physical tasks. From industrial robots on factory floors to service robots in healthcare, robotics is expanding the horizons of automation.

5. Cognitive Automation:
This advanced form of automation leverages AI to handle tasks that require cognitive abilities, such as decision-making, problem-solving, and understanding natural language. It is used in fields like finance, customer service, and supply chain management.

Applications and Impact

AI and automation are transforming industries and everyday life. In healthcare, AI assists in diagnosing diseases, personalizing treatment plans, and managing patient care. In finance, algorithms detect fraudulent transactions, optimize investment strategies, and automate customer support. The retail industry uses AI to manage inventory, predict consumer behavior, and enhance the shopping experience through personalized recommendations.

The impact of these technologies extends to transportation with autonomous vehicles, agriculture with smart farming techniques, and education with personalized learning platforms. They also play a critical role in addressing global challenges, such as climate change, by optimizing energy consumption and improving environmental monitoring.

Challenges and Considerations

Despite their potential, AI and automation pose several challenges. Ethical considerations include biases in AI algorithms, data privacy concerns, and the transparency of AI decision-making processes. Additionally, the potential displacement of jobs due to automation requires careful management to ensure a fair transition for the workforce.

Regulatory frameworks and policies need to evolve to address these challenges, promoting responsible development and deployment of AI and automation technologies. Collaboration between governments, industries, and academia is essential to navigate the complex landscape and harness the benefits while mitigating the risks.

AI and automation are reshaping our world, offering unprecedented opportunities for innovation and efficiency. As these technologies continue to advance, understanding their fundamentals, applications, and implications is vital for individuals, businesses, and society at large. By embracing the potential of AI and automation, we can drive progress and create a future where humans and machines work together to solve the most pressing challenges.

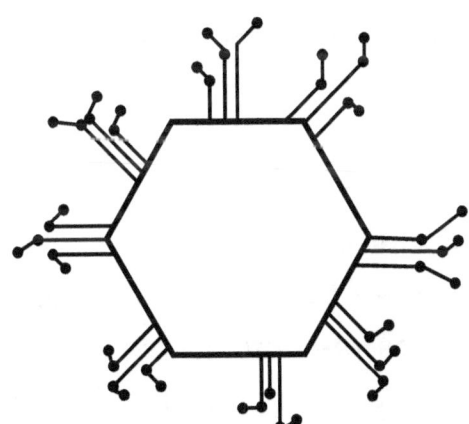

Chapter 2.
The Current Landscape: AI in Today's Workforce

AI and Automation in Various Industries

AI and automation are transforming numerous industries by enhancing efficiency, reducing costs, and enabling innovative solutions. Here are some examples of how these technologies are being applied across different sectors:

Healthcare

1. Medical Imaging and Diagnostics:
AI algorithms analyze medical images such as X-rays, MRIs, and CT scans to detect abnormalities and diagnose conditions with high accuracy. For instance, AI systems can identify early signs of diseases like cancer or neurological disorders, aiding in prompt and accurate diagnosis.

2. Personalized Medicine:
AI analyzes patient data to tailor treatments to individual needs. This includes predicting responses to medications and optimizing treatment plans based on genetic information, lifestyle, and other factors.

3. Robotic Surgery:
Surgical robots, guided by AI, assist surgeons in performing precise and minimally invasive procedures. These robots enhance accuracy, reduce recovery times,

and minimize the risk of complications.

Finance

1. Fraud Detection:
AI systems analyze transaction data to identify unusual patterns and flag potential fraudulent activities. Machine learning models can detect anomalies in real-time, helping to prevent fraud before it occurs.

2. Algorithmic Trading:
AI algorithms execute trades at high speeds and volumes based on predefined criteria. These systems can analyze market data, predict trends, and make investment decisions, often outperforming human traders.

3. Customer Service:
AI-powered chatbots and virtual assistants provide customer support by answering queries, resolving issues, and guiding users through various processes. This reduces the workload on human agents and improves response times.

Manufacturing

1. Predictive Maintenance: AI monitors machinery and equipment to predict when maintenance is needed, reducing downtime and preventing costly breakdowns. Sensors collect data on machine performance, which AI analyzes to identify signs of wear and tear.

2. Quality Control:
Computer vision systems inspect products on production lines to detect defects and ensure quality standards are met. AI can identify minute inconsistencies that might be missed by human inspectors.

3. Supply Chain Optimization:
AI optimizes supply chain operations by forecasting demand, managing inventory, and coordinating logistics. This ensures timely delivery of materials and products while minimizing costs.

Retail

1. Personalized Shopping Experiences:
AI analyzes customer data to provide personalized product recommendations and marketing messages. This enhances customer engagement and drives sales by suggesting items based on individual preferences and browsing history.

2. Inventory Management:
Automation systems track inventory levels in real-time, automatically reordering stock when necessary. AI algorithms predict future demand, helping retailers manage their supply chain more efficiently.

3. Customer Service and Support:
AI chatbots assist customers with inquiries, returns, and other service issues, providing instant support and freeing

up human staff for more complex tasks.

Agriculture

1. Precision Farming:
AI-powered systems monitor soil conditions, weather patterns, and crop health to optimize farming practices. Drones equipped with sensors and cameras collect data, which AI analyzes to guide planting, irrigation, and harvesting decisions.

2. Automated Machinery:
Autonomous tractors and harvesters perform agricultural tasks with minimal human intervention, increasing efficiency and reducing labor costs. These machines use AI to navigate fields and optimize operations.

3. Pest and Disease Control:
AI detects early signs of pest infestations and plant diseases, allowing for timely intervention. This helps farmers minimize crop damage and reduce the use of pesticides.

Transportation

1. Autonomous Vehicles:
Self-driving cars and trucks use AI to navigate roads, interpret traffic signals, and avoid obstacles. These vehicles have the potential to reduce accidents, improve

traffic flow, and lower transportation costs.

2. Traffic Management:
AI systems analyze traffic data to optimize traffic light timings and manage congestion. Real-time adjustments based on traffic patterns can improve overall traffic efficiency and reduce delays.

3. Fleet Management:
AI optimizes routes for delivery trucks and other fleet vehicles, reducing fuel consumption and improving delivery times. It also monitors vehicle performance to schedule maintenance and prevent breakdowns.

Entertainment

1. Content Recommendations:
Streaming services like Netflix and Spotify use AI to recommend movies, TV shows, and music based on user preferences. These personalized suggestions enhance user experience and engagement.

2. Content Creation:
AI assists in generating content, such as music compositions, news articles, and even screenplays. This technology can augment creative processes and produce new forms of entertainment.

3. Gaming:
AI creates intelligent non-player characters (NPCs) that

provide realistic and challenging interactions for players. It also personalizes gaming experiences by adapting to individual playing styles.

These examples illustrate the broad and profound impact of AI and automation across various industries. As these technologies continue to advance, their applications will likely expand, driving further innovation and efficiency improvements.

Case studies of companies implementing AI

The implementation of AI across various industries has led to significant advancements in efficiency, innovation, and customer satisfaction. Here are several case studies that highlight how companies are leveraging AI to transform their operations and achieve remarkable outcomes.

1. Google: Enhancing Search and Advertising

Google is at the forefront of AI innovation, utilizing machine learning and deep learning to enhance its core products and services.

Search Algorithms:
Google uses AI to improve its search algorithms, providing more accurate and relevant search results. The RankBrain algorithm, an AI-based system, helps process and understand search queries, particularly those that are

complex or ambiguous. This leads to better user experiences and more precise answers.

Advertising:
AI powers Google Ads, optimizing ad placements and targeting to maximize click-through rates and conversions. Machine learning algorithms analyze user behavior, preferences, and past interactions to serve personalized ads that are more likely to engage users.

2. Amazon: Revolutionizing E-Commerce and Logistics

Amazon employs AI across multiple facets of its business, from customer recommendations to warehouse automation.

Recommendation Engine:
Amazon's recommendation system uses AI to analyze customer data and suggest products based on browsing history, purchase behavior, and preferences. This personalized approach drives sales and enhances customer satisfaction.

Warehouse Automation:
Amazon's fulfillment centers are equipped with AI-driven robots that assist in picking, packing, and sorting items. These robots work alongside human employees to increase efficiency and speed up order processing. The use of AI in logistics helps Amazon manage its vast inventory and ensure timely deliveries

3. Tesla: Pioneering Autonomous Driving

Tesla is a leader in the development and deployment of AI for autonomous driving technology.

Autopilot System:
Tesla's Autopilot and Full Self-Driving (FSD) systems rely on AI to navigate roads, interpret traffic signals, and avoid obstacles. The AI uses data from cameras, sensors, and radar to make real-time driving decisions. Continuous learning from a fleet of vehicles helps improve the system's performance and safety over time.

Manufacturing:
AI also plays a crucial role in Tesla's manufacturing process. AI-driven robots perform complex tasks on the assembly line, ensuring precision and quality in vehicle production. This automation reduces production costs and increases output.

4. IBM: Transforming Healthcare with Watson

IBM Watson is an AI system that has made significant strides in the healthcare industry.

Clinical Decision Support:
Watson for Oncology helps oncologists diagnose and treat cancer by analyzing large volumes of medical literature, patient records, and clinical trial data. It provides evidence-based treatment recommendations tailored to

individual patients, improving the accuracy and effectiveness of care.

Drug Discovery:
IBM Watson is also used in drug discovery, where it accelerates the research process by identifying potential drug candidates and predicting their efficacy. This application of AI helps pharmaceutical companies bring new treatments to market faster.

5. Netflix: Personalizing Entertainment

Netflix utilizes AI to enhance its streaming service, ensuring users have a personalized and enjoyable viewing experience.

Content Recommendations:
Netflix's recommendation engine uses AI to analyze user preferences, viewing history, and ratings to suggest movies and TV shows that match individual tastes. This personalization keeps users engaged and encourages longer viewing times.

Content Creation:
AI assists Netflix in content creation by analyzing trends and viewer data to predict the success of new shows and movies. This helps Netflix make informed decisions about which projects to greenlight, ensuring a steady stream of popular content.

6. Microsoft: Enhancing Productivity with AI

Microsoft integrates AI into its products to boost productivity and provide smarter solutions for businesses and individuals.

Office 365:
AI features in Microsoft Office 365, such as the intelligent writing assistant in Word and Excel's data insights, help users work more efficiently. AI-powered tools provide grammar suggestions, data visualization, and automated email sorting.

Azure AI:
Microsoft Azure offers AI services that enable businesses to build and deploy AI models at scale. Companies use Azure AI to develop applications for customer service, predictive maintenance, and fraud detection, leveraging Microsoft's robust AI infrastructure.

7. Uber: Optimizing Ride-Hailing Services

Uber employs AI to improve its ride-hailing services, making them more efficient and user-friendly.

Route Optimization:
AI algorithms analyze traffic patterns, historical data, and real-time conditions to optimize routes for drivers. This reduces travel time, fuel consumption, and costs, enhancing the overall efficiency of the service.

Dynamic Pricing:
Uber uses AI for dynamic pricing, adjusting fares based on demand, traffic conditions, and availability of drivers. This ensures a balance between supply and demand, providing timely rides for customers and fair earnings for drivers.

These case studies illustrate the diverse applications of AI across various industries. Companies like Google, Amazon, Tesla, IBM, Netflix, Microsoft, and Uber are leading the way in leveraging AI to enhance their products, services, and operations, setting the stage for a future where AI plays a central role in driving innovation and efficiency.

The State of the Job Market

The job market is currently experiencing a dynamic shift, influenced by technological advancements, economic changes, and evolving workforce demands. Understanding the state of the job market today requires examining several key trends and factors that are shaping employment opportunities and challenges.

Impact of Technological Advancements

1. Automation and AI:
Automation and artificial intelligence are significantly transforming the job landscape. While these technologies are enhancing productivity and efficiency, they are also leading to the displacement of certain jobs, particularly those involving routine and

repetitive tasks. However, they are simultaneously creating new job opportunities in fields such as AI development, data analysis, and cybersecurity.

2. Digital Transformation:

The rise of digital technologies is driving demand for skills in software development, digital marketing, and IT support. Businesses across various sectors are investing in digital infrastructure, leading to a surge in job openings for tech-savvy professionals who can help navigate and implement these changes.

Economic Factors

1. Post-Pandemic Recovery:

The COVID-19 pandemic has had a lasting impact on the job market. While some industries, like hospitality and travel, are still recovering, others, such as healthcare and e-commerce, have seen significant growth. The pandemic accelerated remote work adoption, which has become a permanent fixture for many organizations, influencing job availability and location preferences.

2. Globalization:

Global economic integration continues to influence the job market. Companies are increasingly looking for talent across borders, facilitated by remote work technologies. This has expanded opportunities for workers globally but also intensified competition for certain roles.

Evolving Workforce Demands

1. Skill Gaps:
There is a growing emphasis on skills over traditional qualifications. Employers are seeking candidates with specific technical skills, such as programming languages, data analytics, and cloud computing, as well as soft skills like adaptability, problem-solving, and communication. Continuous learning and upskilling are becoming essential for career advancement.

2. Gig Economy:
The gig economy is expanding, with more individuals opting for freelance, part-time, or contract work. Platforms like Uber, Upwork, and Fiverr are facilitating this shift, offering flexibility and diverse income streams. However, this trend also raises concerns about job security, benefits, and workers' rights.

Industry-Specific Trends

1. Healthcare:
The healthcare industry is experiencing robust growth, driven by an aging population and the increased focus on health and wellness. There is high demand for healthcare professionals, including nurses, doctors, and allied health workers, as well as roles in healthcare administration and telemedicine.

2. Technology:

The tech industry remains a powerhouse of job creation, with ongoing demand for software developers, cybersecurity experts, data scientists, and AI specialists. Innovations in fields like artificial intelligence, blockchain, and quantum computing are further expanding the scope of tech employment.

3. Renewable Energy:

The shift towards sustainable energy sources is creating new job opportunities in the renewable energy sector. Jobs in solar and wind energy, energy storage, and electric vehicle manufacturing are on the rise as governments and companies commit to reducing carbon footprints and investing in green technologies.

Challenges and Opportunities

1. Inequality and Access:

Despite the overall positive trends, there are challenges related to inequality and access to opportunities. Economic disparities, regional differences, and barriers to education and training can limit access to high-quality jobs for certain groups. Addressing these issues requires targeted policies and initiatives to promote inclusive growth.

2. Work-Life Balance:

The increase in remote work has brought attention to the importance of work-life balance. While remote work offers flexibility, it also blurs the boundaries between personal and professional life. Companies are exploring ways to support employees' well-being through flexible schedules, mental health resources, and a focus on work-life integration.

The state of the job market is characterized by rapid change and adaptation. Technological advancements, economic shifts, and evolving workforce demands are reshaping how and where people work. While challenges such as job displacement and inequality persist, opportunities abound for those who can adapt and acquire the skills needed in this new landscape. Understanding these trends and preparing for continuous learning will be crucial for navigating the future of work successfully.

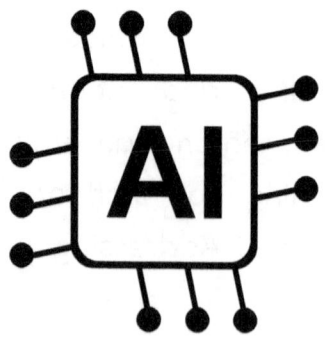

Chapter 3.
Automation in Action: Transforming Industries

Manufacturing and robotics

Manufacturing and robotics are two interconnected fields that are revolutionizing the industrial landscape. The integration of robotics into manufacturing processes is enhancing efficiency, precision, and flexibility, driving the industry toward greater innovation and productivity.

The Role of Robotics in Manufacturing

1. Automation of Repetitive Tasks:
Robots excel at performing repetitive, monotonous tasks with high precision and consistency. In manufacturing, this includes activities such as assembly, welding, painting, and packaging. By automating these processes, companies can increase output while maintaining consistent quality, reducing human error, and lowering labor costs.

2. Advanced Robotics:
Modern manufacturing robots are equipped with advanced sensors, machine learning algorithms, and AI capabilities. These robots can adapt to new tasks, work collaboratively with human workers, and make real-time decisions. Collaborative robots, or cobots, are designed to work alongside humans, enhancing productivity without replacing the human workforce.

Benefits of Robotics in Manufacturing

1. Increased Efficiency and Productivity:
Robotics can operate 24/7 without fatigue, significantly boosting production rates and efficiency. This continuous operation helps manufacturers meet high demand and reduces lead times.

2. Enhanced Precision and Quality:
Robots perform tasks with high accuracy, ensuring uniformity and quality in products. This precision is critical in industries such as electronics, automotive, and pharmaceuticals, where even minor defects can have significant consequences.

3. Safety and Ergonomics:
Robots handle hazardous and strenuous tasks, improving workplace safety and reducing the risk of injuries. This allows human workers to focus on more complex, value-added activities and enhances overall job satisfaction.

Applications of Robotics in Manufacturing

1. Automotive Industry:
The automotive sector has been a pioneer in adopting robotics. Robots are used extensively for welding, painting, assembly, and quality inspection. They help in producing high-quality vehicles at a faster rate and lower cost.

2. Electronics Manufacturing:
The electronics industry uses robots for tasks such as circuit board assembly, soldering, and packaging. The precision and speed of robots are crucial for handling small components and maintaining high production volumes.

3. Food and Beverage:
Robotics in food and beverage manufacturing include tasks like sorting, packaging, and palletizing. Robots ensure hygiene and consistency, which are vital in food processing.

4. Pharmaceuticals:
In the pharmaceutical industry, robots assist in drug manufacturing, packaging, and quality control. They help maintain sterile environments and ensure the precise handling of sensitive materials.

Challenges and Future Directions

1. Integration and Adaptation:
Integrating robotics into existing manufacturing systems can be complex and costly. It requires careful planning and customization to ensure compatibility with current processes and infrastructure.

2. Workforce Transition:
As robotics automate more tasks, there is a need for workforce reskilling and upskilling. Workers must be trained to manage and maintain robotic systems, and there should

be a focus on transitioning employees to roles that require human creativity and problem-solving.

3. Technological Advancements:
The future of manufacturing robotics lies in continued advancements in AI, machine learning, and sensor technology. These innovations will enable robots to become more autonomous, adaptable, and capable of handling increasingly complex tasks.

4. Sustainability:
Robotics can contribute to sustainable manufacturing by optimizing resource use, reducing waste, and improving energy efficiency. The development of eco-friendly robotic technologies and practices is becoming a priority for many manufacturers.

Robotics is transforming manufacturing by enhancing efficiency, precision, and safety. While the integration of robotics presents challenges, the benefits far outweigh the drawbacks, leading to increased productivity and innovation. As technology advances, the role of robotics in manufacturing will continue to expand, driving the industry toward a more automated, efficient, and sustainable future. Understanding and embracing these changes will be crucial for manufacturers looking to remain competitive in the global market.

AI in Healthcare

Artificial Intelligence (AI) is revolutionizing the healthcare industry by enhancing diagnostics, personalizing treatment plans, and improving operational efficiency. The integration of AI into healthcare is creating new opportunities for better patient outcomes, streamlined processes, and cost savings.

Applications of AI in Healthcare

1. Medical Imaging and Diagnostics:
AI algorithms analyze medical images such as X-rays, MRIs, and CT scans to identify abnormalities with high accuracy. AI systems can detect early signs of diseases like cancer, neurological disorders, and cardiovascular conditions, often outperforming human radiologists in speed and precision. This leads to earlier diagnosis and more effective treatment plans.

2. Predictive Analytics:
AI uses data from electronic health records (EHRs), genetic information, and other sources to predict patient outcomes and disease progression. Predictive analytics can identify patients at high risk for conditions such as diabetes, heart disease, and sepsis, enabling proactive interventions that can prevent complications and reduce hospital readmissions.

3. Personalized Medicine:

AI helps tailor treatments to individual patients based on their genetic makeup, lifestyle, and medical history. By analyzing vast amounts of data, AI can recommend personalized treatment plans and predict how patients will respond to specific medications, enhancing the effectiveness of therapies and minimizing adverse effects.

4. Drug Discovery and Development:

AI accelerates the drug discovery process by identifying potential drug candidates and predicting their efficacy and safety profiles. Machine learning algorithms analyze biological data to uncover new therapeutic targets and optimize clinical trial designs, reducing the time and cost associated with bringing new drugs to market.

5. Virtual Health Assistants:

AI-powered virtual assistants provide patients with medical advice, reminders for medication adherence, and answers to health-related questions. These virtual assistants can also assist healthcare providers by managing administrative tasks, scheduling appointments, and maintaining patient records, improving the overall efficiency of healthcare delivery.

Benefits of AI in Healthcare

1. Improved Accuracy and Speed:

AI systems can process and analyze large volumes of data quickly and accurately, leading to faster and more reliable diagnoses. This enhances the ability of healthcare providers to make informed decisions and deliver timely care.

2. Enhanced Patient Care:
AI-driven personalized medicine and predictive analytics enable healthcare providers to offer more targeted and effective treatments. This leads to better patient outcomes, reduced side effects, and improved quality of life.

3. Operational Efficiency:
AI streamlines administrative and clinical workflows, reducing the burden on healthcare staff. Automation of routine tasks such as data entry, appointment scheduling, and billing allows healthcare professionals to focus on patient care, increasing overall efficiency and productivity.

4. Cost Savings:
By improving diagnostic accuracy, optimizing treatment plans, and preventing disease progression, AI can reduce healthcare costs. Early detection and intervention can prevent costly hospitalizations and treatments, while operational efficiencies lower administrative expenses.

Challenges and Considerations

1. Data Privacy and Security:
The use of AI in healthcare involves handling sensitive

patient data, raising concerns about privacy and security. Ensuring robust data protection measures and compliance with regulations such as HIPAA (Health Insurance Portability and Accountability Act) is essential to maintain patient trust and confidentiality.

2. Integration with Existing Systems:
Integrating AI technologies with existing healthcare systems and workflows can be challenging. Healthcare providers must ensure that AI tools are compatible with their current infrastructure and that staff are adequately trained to use these technologies effectively.

3. Ethical and Bias Issues:
AI algorithms can inherit biases present in the data they are trained on, potentially leading to biased outcomes. It is crucial to develop and implement AI systems that are transparent, fair, and free from biases to ensure equitable healthcare for all patients.

4. Regulatory and Legal Challenges:
The use of AI in healthcare is subject to regulatory scrutiny to ensure safety and efficacy. Navigating the complex regulatory landscape and obtaining necessary approvals can be a lengthy and challenging process for AI developers and healthcare providers.

Future Directions

The future of AI in healthcare holds immense potential for

further advancements and innovations. Continued research and development in AI technologies will lead to more sophisticated and effective applications, enhancing every aspect of healthcare delivery. Key areas of focus include:

Integration of AI with IoT:
Combining AI with Internet of Things (IoT) devices, such as wearable health monitors, will provide real-time health data and enable continuous monitoring of patients, leading to more proactive and personalized care.

AI-Driven Clinical Decision Support:
AI will increasingly support healthcare providers in making complex clinical decisions by analyzing patient data and providing evidence-based recommendations.

Expansion of Telemedicine:
AI will enhance telemedicine services by enabling remote diagnostics, virtual consultations, and personalized treatment plans, making healthcare more accessible and convenient for patients.

AI is transforming healthcare by improving diagnostic accuracy, personalizing treatments, and enhancing operational efficiency. While challenges such as data privacy, integration, and ethical considerations must be addressed, the potential benefits of AI in healthcare are immense. As AI technologies continue to evolve, they will play a crucial role in shaping the future of healthcare.

The Impact on Finance and Banking

Artificial intelligence (AI) and automation are significantly transforming the finance and banking sectors, enhancing operational efficiency, customer experience, risk management, and decision-making processes. These technologies are reshaping how financial institutions operate and interact with customers, offering both opportunities and challenges.

Enhancing Operational Efficiency

1. Automated Processes:
AI-powered automation streamlines routine and repetitive tasks such as data entry, transaction processing, and compliance reporting. This reduces operational costs, minimizes human error, and frees up employees to focus on higher-value activities. For example, robotic process automation (RPA) can handle back-office tasks like loan processing and account management efficiently and accurately.

2. Customer Service:
AI-driven chatbots and virtual assistants provide round-the-clock customer support, handling inquiries, resolving issues, and assisting with transactions. These tools enhance customer satisfaction by providing immediate and accurate responses while reducing the workload on human customer service representatives.

Improving Risk Management

1. Fraud Detection and Prevention:

AI systems analyze transaction data in real-time to detect unusual patterns and potential fraudulent activities. Machine learning algorithms can identify anomalies that may indicate fraud, enabling financial institutions to act swiftly to prevent losses. AI's ability to continuously learn and adapt makes it increasingly effective at identifying new types of fraud.

2. Credit Scoring and Risk Assessment:

AI enhances the accuracy of credit scoring models by incorporating a broader range of data sources, including non-traditional data such as social media activity and payment history from utilities. This allows for more precise risk assessment and better decision-making in lending processes. AI models can also predict the likelihood of default, helping banks manage their credit portfolios more effectively.

Enhancing Customer Experience

1. Personalized Financial Services:

AI analyzes customer data to offer personalized financial advice and product recommendations. By understanding individual customer needs and behaviors, banks can tailor their services to provide more relevant and timely offers. For example, AI-driven investment platforms can create customized investment portfolios based on a user's risk tolerance and financial goals.

2. Streamlined Onboarding:
AI simplifies the customer onboarding process through automated identity verification and document processing. This reduces the time required to open new accounts and enhances the overall customer experience. AI can also ensure compliance with regulatory requirements, such as anti-money laundering (AML) and know-your-customer (KYC) procedures, by efficiently verifying customer information.

Advancing Decision-Making

1. Data-Driven Insights:
AI and machine learning algorithms analyze vast amounts of financial data to uncover patterns, trends, and insights that inform strategic decisions. Financial institutions use these insights for market analysis, investment strategies, and identifying new business opportunities. Predictive analytics helps banks forecast market movements and adjust their strategies accordingly.

2. Algorithmic Trading:
AI plays a crucial role in algorithmic trading, where it processes market data and executes trades at high speeds and volumes based on predefined criteria. AI algorithms can analyze multiple data sources, predict market trends, and make investment decisions faster and more accurately than human traders, potentially leading

to higher returns.o safeguard customer information.

Challenges and Considerations

1. Data Privacy and Security:
The use of AI in finance involves handling sensitive customer data, raising concerns about privacy and security. Financial institutions must implement robust data protection measures and comply with regulations such as the General Data Protection Regulation (GDPR) t

2. Ethical and Bias Issues:
AI algorithms can inherit biases present in the data they are trained on, leading to biased outcomes in lending, investment, and other financial decisions. It is essential to develop transparent and fair AI systems to ensure ethical use and avoid discrimination.

3. Regulatory Compliance:
The integration of AI in finance must align with regulatory frameworks to ensure compliance. Regulatory bodies are increasingly scrutinizing AI applications to ensure they meet legal and ethical standards. Financial institutions must navigate these regulations while leveraging AI's benefits.

Future Directions

1. Continued Innovation:
The finance and banking sectors will continue to innovate

with AI, exploring new applications such as decentralized finance (DeFi) and blockchain technology. AI will play a crucial role in enhancing the security, efficiency, and transparency of these emerging financial ecosystems.

2. Enhanced Human-AI Collaboration:

The future will likely see more collaborative interactions between humans and AI, where AI tools augment human capabilities rather than replace them. Financial professionals will need to develop skills to work alongside AI systems, leveraging their strengths to make more informed and strategic decisions.

3. Focus on Sustainability:

AI can help financial institutions assess and manage environmental, social, and governance (ESG) risks, promoting sustainable investment practices. By analyzing ESG data, AI can identify companies with strong sustainability profiles, guiding investment strategies that align with ethical and environmental goals.

AI is profoundly impacting the finance and banking sectors by enhancing operational efficiency, improving risk management, personalizing customer experiences, and advancing decision-making processes. While there are challenges related to data privacy, ethics, and regulatory compliance, the potential benefits of AI in finance are immense. As AI technology continues to evolve, it will further transform the financial landscape, offering new opportunities for innovation and growth. Financial institutions that embrace AI and navigate its challenges effectively will be well-positioned to thrive in the future.

Agriculture and Smart Farming

Agriculture and smart farming are undergoing a revolutionary transformation driven by the integration of advanced technologies such as the Internet of Things (IoT), artificial intelligence (AI), and robotics. These innovations are enhancing productivity, sustainability, and efficiency in the agricultural sector, addressing the growing global demand for food while minimizing environmental impact.

The Role of Technology in Smart Farming

1. Internet of Things (IoT):
IoT devices, including sensors, drones, and connected machinery, collect real-time data on various aspects of farming. These devices monitor soil conditions, weather patterns, crop health, and livestock behavior, providing farmers with valuable insights to make informed decisions. For instance, soil sensors measure moisture levels, enabling precise irrigation that conserves water while maintaining optimal soil conditions for plant growth.

2. Artificial Intelligence (AI):
AI algorithms analyze the data gathered by IoT devices to provide actionable recommendations and automate

decision-making processes. AI systems can predict weather changes, identify crop diseases, and optimize planting and harvesting schedules. For example, AI-powered platforms can analyze satellite imagery to detect early signs of pest infestations or nutrient deficiencies, allowing farmers to take corrective actions promptly.

3. Robotics and Automation:

Robots and automated machinery perform a variety of agricultural tasks, such as planting, harvesting, weeding, and sorting. These technologies increase efficiency and reduce labor costs. Autonomous tractors, for example, can operate continuously without human intervention, even in challenging weather conditions. Robots equipped with advanced sensors and AI can identify and remove weeds with precision, reducing the need for chemical herbicides.

Benefits of Smart Farming

1. Increased Productivity:

Smart farming technologies enable farmers to maximize crop yields and livestock production. Precision farming techniques ensure that resources such as water, fertilizers, and pesticides are used efficiently, leading to healthier crops and higher outputs. Automated machinery and AI-driven management systems streamline operations, reducing downtime and increasing overall farm productivity.

2. Sustainability and Environmental Impact:

Smart farming promotes sustainable agricultural practices by minimizing resource wastage and reducing the environmental footprint. Precision irrigation systems deliver the right amount of water to plants, conserving water resources. Similarly, targeted application of fertilizers and pesticides reduces chemical runoff and soil degradation, protecting ecosystems and promoting biodiversity.

3. Cost Efficiency:

By optimizing resource use and automating labor-intensive tasks, smart farming reduces operational costs. Farmers can save on inputs like water, fertilizers, and labor while achieving better yields. Additionally, early detection of diseases and pests through AI systems prevents crop losses, further enhancing profitability.

4. Improved Decision-Making:

Data-driven insights from IoT devices and AI systems enable farmers to make more informed decisions. Predictive analytics help anticipate weather events, market trends, and potential risks, allowing farmers to plan and adapt proactively. This leads to more resilient farming practices and better financial outcomes.

Applications of Smart Farming

1. Precision Agriculture:

Precision agriculture involves using technology to monitor and manage crop production at a granular level.

GPS-guided equipment, soil sensors, and drones collect data on field conditions, enabling farmers to apply inputs like water and nutrients precisely where needed. This enhances crop health and yield while conserving resources.

2. Livestock Monitoring:

IoT devices and AI are used to monitor livestock health and behavior. Wearable sensors track vital signs, activity levels, and feeding patterns, alerting farmers to any signs of illness or distress. AI systems analyze this data to optimize feeding schedules, detect health issues early, and improve overall herd management.

3. Automated Greenhouses:

Smart greenhouses use IoT sensors and AI to control environmental factors such as temperature, humidity, and light. Automated systems adjust these conditions to create the optimal environment for plant growth, enhancing yield and reducing the need for manual intervention.

4. Supply Chain Optimization:

AI and IoT technologies improve supply chain management by providing real-time visibility into inventory levels, logistics, and market demand. This helps farmers and distributors manage resources more efficiently, reduce waste, and ensure timely delivery of fresh produce.

Challenges and Future Directions

1. Data Management and Privacy:
The widespread use of IoT and AI in agriculture generates vast amounts of data. Managing this data effectively and ensuring its privacy and security are significant challenges. Developing robust data management systems and adhering to data privacy regulations are crucial for the successful implementation of smart farming technologies.

2. High Initial Costs:
The adoption of smart farming technologies often requires significant initial investment. While these technologies can lead to cost savings and increased productivity in the long run, the upfront costs can be a barrier for small and medium-sized farms. Financial support and incentives from governments and organizations can help mitigate this challenge.

3. Technical Expertise:
The effective use of smart farming technologies requires technical knowledge and skills. Farmers need training and support to understand and implement these technologies successfully. Building partnerships with technology providers and educational institutions can facilitate the transfer of knowledge and skills to the farming community.

4. Sustainability and Ethics:
While smart farming promotes sustainability, it also raises ethical questions about the use of AI and automation in

agriculture. Ensuring that these technologies are used responsibly and equitably is essential for maintaining trust and achieving sustainable agricultural practices.

Smart farming represents the future of agriculture, offering numerous benefits in terms of productivity, sustainability, and cost efficiency. The integration of IoT, AI, and robotics is transforming traditional farming practices, enabling farmers to meet the growing global demand for food while minimizing environmental impact. Despite the challenges, the continued advancement and adoption of smart farming technologies hold great promise for a more sustainable and efficient agricultural sector.

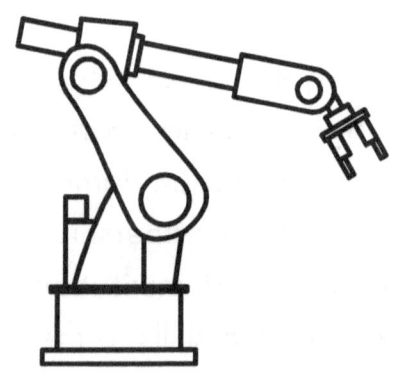

Chapter 4.
Skills for the Future: Education and Training

Identifying future-proof skills

In an era of rapid technological advancement and evolving job markets, identifying and cultivating future-proof skills is crucial for career resilience and success. Future-proof skills are those that remain valuable and relevant despite changes in industry landscapes, ensuring individuals can adapt to new roles and opportunities as they arise. Here are key areas where future-proof skills can be developed:

1. Digital Literacy and Technology Skills

Basic Digital Literacy:
Understanding fundamental digital tools and platforms is essential. Proficiency in using word processors, spreadsheets, email, and other basic software forms the foundation of digital literacy.

Advanced Technology Skills:
As technology evolves, advanced skills such as coding, data analysis, and cybersecurity become increasingly important. Familiarity with emerging technologies like artificial intelligence, machine learning, and blockchain can provide a competitive edge.

2. Critical Thinking and Problem-Solving

Analytical Thinking:
The ability to analyze data, identify trends, and draw meaningful conclusions is crucial. Analytical thinkers can approach problems methodically, breaking them down into manageable parts and devising effective solutions.

Creative Problem-Solving:
Creativity in problem-solving involves thinking outside the box and developing innovative solutions. This skill is valuable in environments that require adaptability and innovation.

3. Emotional Intelligence (EQ)

Self-Awareness:
Understanding one's emotions, strengths, and weaknesses is key to personal and professional growth. Self-aware individuals can manage their emotions effectively and make informed decisions.

Empathy and Relationship Management:
The ability to empathize with others and manage relationships is critical in team-based and customer-facing roles. High EQ enables better communication, conflict resolution, and collaboration.

4. Adaptability and Learning Agility

Lifelong Learning:
Commitment to continuous learning and skill development is essential in a rapidly changing job market. Individuals who embrace lifelong learning can quickly acquire new skills and knowledge, staying relevant in their careers.

Adaptability:
Being open to change and able to adjust to new circumstances is vital. Adaptable individuals can navigate transitions smoothly and remain productive in different environments.

5. Communication and Collaboration

Effective Communication:
Clear and concise communication is important in all professional settings. This includes verbal, written, and digital communication skills, ensuring that ideas and information are conveyed accurately and persuasively.

Team Collaboration:
The ability to work effectively in teams is crucial as workplaces become more collaborative. Skills in teamwork, including active listening, cooperation, and shared problem-solving, enhance productivity and innovation.

6. Leadership and Management

Leadership Skills:
Leadership involves inspiring and guiding others toward achieving common goals. Key leadership skills include decision-making, strategic thinking, and the ability to motivate and empower teams.

Project Management:
Managing projects efficiently requires planning, organization, and time management skills. Project managers must coordinate resources, set goals, and ensure projects are completed on time and within budget.

7. Specialized Knowledge and Expertise

Domain-Specific Skills:
Specialized knowledge in a particular field or industry can provide a significant advantage. Staying updated with the latest trends and advancements in one's area of expertise ensures continued relevance and value.

Interdisciplinary Knowledge:
Combining expertise from different disciplines can lead to innovative solutions and new opportunities. Interdisciplinary knowledge allows individuals to bridge gaps between fields and apply diverse perspectives to complex problems.

Strategies for Developing Future-Proof Skills

Continuous Education:
Enroll in courses, attend workshops, and pursue certifications to stay updated with new skills and knowledge. Online platforms and educational institutions offer a wide range of opportunities for continuous learning.

Networking and Mentorship:
Engage with professionals in your field to gain insights and advice. Networking and mentorship provide valuable opportunities for learning and career growth.

Practical Experience:
Apply new skills through hands-on experience. Internships, volunteer work, and project-based learning can help solidify theoretical knowledge and develop practical expertise.

Self-Assessment and Feedback:
Regularly assess your skills and seek feedback from peers and supervisors. Understanding your strengths and areas for improvement helps guide your professional development efforts.

Identifying and developing future-proof skills is essential for navigating the dynamic job market and ensuring long-term career success. By focusing on digital literacy, critical thinking, emotional intelligence, adaptability, communication, leadership, and specialized knowledge, individuals can build a robust skill set that remains valuable in the face of technological advancements and industry changes. Embracing continuous learning and practical experience will further enhance these skills, preparing professionals for the challenges and opportunities of the future.

The Role of Education Systems

Education systems play a pivotal role in preparing individuals for the demands of the modern workforce and fostering lifelong learning. In an era marked by rapid technological advancements and shifting job markets, education must evolve to equip students with the skills and knowledge needed to thrive in an increasingly complex and dynamic world. Here are key aspects of the role of education systems:

1. Developing Foundational Skills

Literacy and Numeracy:
At the core of any education system is the commitment to teaching basic literacy and numeracy. These foundational skills are essential for further learning and effective participation in society.

Critical Thinking and Problem-Solving:
Education systems must emphasize the development of critical thinking and problem-solving abilities. These skills

enable students to analyze information, make informed decisions, and address complex challenges creatively.

2. Promoting Digital Literacy

Basic Digital Skills:
In today's digital age, proficiency with digital tools and technologies is crucial. Education systems should ensure that students are comfortable using computers, software applications, and the internet from an early age.

Advanced Technology Skills:
Beyond basic digital literacy, education should offer training in more advanced areas such as coding, data analysis, and cybersecurity. Familiarity with emerging technologies like artificial intelligence and machine learning can provide students with a competitive edge in the job market.

3. Encouraging Lifelong Learning

Cultivating a Growth Mindset:
Education systems should foster a culture of lifelong learning by encouraging a growth mindset. This involves instilling a belief in the value of continuous learning and the ability to adapt and grow throughout one's life.

Access to Continuous Education:
Providing opportunities for ongoing education and professional development is essential. This can include

adult education programs, online courses, and vocational training that allow individuals to update their skills and knowledge regularly.

4. Fostering Emotional Intelligence

Self-Awareness and Self-Management:
Emotional intelligence (EQ) is increasingly recognized as critical for personal and professional success. Education systems should include programs that help students develop self-awareness and self-management skills.

Social Skills and Empathy:
Teaching students to understand and manage their emotions, communicate effectively, and empathize with others is vital. These skills are essential for teamwork, leadership, and building healthy relationships.

5. Promoting Inclusivity and Equity

Equal Access to Education:
Ensuring that all students, regardless of their background, have access to quality education is fundamental. Education systems must work to eliminate barriers to education and provide support for disadvantaged and marginalized groups.

Inclusive Curriculum:
An inclusive curriculum that reflects diverse perspectives and experiences helps students develop a broader

understanding of the world and fosters an inclusive mindset.

6. Preparing for the Future Workforce

Career and Technical Education:
Education systems should offer career and technical education (CTE) programs that prepare students for specific careers and trades. These programs provide practical skills and hands-on experience, bridging the gap between education and employment.

Soft Skills Development:
In addition to technical skills, education systems should focus on developing soft skills such as communication, teamwork, adaptability, and leadership. These skills are highly valued by employers and essential for career success.

7. Integrating Interdisciplinary Learning

Holistic Education:
Integrating interdisciplinary learning approaches helps students make connections between different subjects and apply their knowledge to real-world problems. This holistic approach fosters creativity and innovation.

STEM and Beyond:
While STEM (science, technology, engineering, and mathematics) education is crucial, it is also important to

balance it with arts, humanities, and social sciences. This ensures a well-rounded education that prepares students for a variety of career paths.

Strategies for Evolving Education Systems

Curriculum Reform:
Regularly updating and reforming the curriculum to reflect current knowledge, skills, and societal needs is essential. This ensures that education remains relevant and prepares students for the future.

Teacher Training and Support:
Providing ongoing professional development for teachers is crucial for effective education delivery. Teachers need to be equipped with the latest pedagogical techniques and knowledge to inspire and guide students.

Leveraging Technology:
Utilizing technology to enhance learning experiences can make education more engaging and accessible. This includes incorporating digital tools, online resources, and interactive learning platforms.

Community and Industry Partnerships:
Collaborating with community organizations and industries can provide students with real-world learning opportunities and exposure to various career paths. Partnerships can enhance vocational training and ensure that education aligns with workforce demands.

Education systems play a crucial role in shaping the future by equipping individuals with the skills and knowledge needed to navigate an ever-changing world. By focusing on foundational skills, digital literacy, lifelong learning, emotional intelligence, inclusivity, workforce preparation, and interdisciplinary learning, education systems can create a robust framework for personal and professional development. Adapting to the needs of the modern world through curriculum reform, teacher support, technology integration, and community partnerships ensures that education remains relevant, equitable, and capable of preparing students for the challenges and opportunities of the future.

Lifelong Learning and Reskilling

Lifelong learning and reskilling are essential strategies for individuals to remain relevant and competitive in the rapidly evolving job market. As technological advancements and shifting economic landscapes continuously reshape industries, the ability to adapt and acquire new skills throughout one's career has become more critical than ever.

The Importance of Lifelong Learning

1. Continuous Skill Development:
Lifelong learning involves the ongoing pursuit of knowledge and skills, whether for professional development, personal enrichment, or both. This

continuous process helps individuals stay updated with the latest industry trends, technologies, and best practices, ensuring they remain valuable assets in the workforce.

2. Adaptability and Flexibility:
The modern job market is characterized by frequent changes and disruptions. Lifelong learning equips individuals with the ability to adapt to new roles, industries, or technologies. It fosters a mindset of flexibility, enabling workers to pivot their careers in response to changing demands.

3. Personal Growth and Fulfillment:
Beyond professional benefits, lifelong learning contributes to personal growth and fulfillment. Engaging in learning activities can enhance critical thinking, creativity, and problem-solving skills, leading to a more enriched and meaningful life.

The Necessity of Reskilling

1. Technological Advancements:
As automation, AI, and other technologies transform various industries, many traditional jobs are becoming obsolete, while new ones are emerging. Reskilling allows workers to acquire the competencies needed to transition into these new roles, ensuring they remain employable.

2. Economic Shifts:
Global economic shifts, such as the transition to a green

economy or the rise of the gig economy, demand new skill sets. Reskilling helps workers align their abilities with the requirements of these emerging economic sectors, enhancing their career prospects.

3. Job Security:
By continuously updating their skills, workers can improve their job security. Employers value employees who demonstrate a commitment to learning and adaptability, making them more likely to retain and promote such individuals.

Key Areas for Lifelong Learning and Reskilling

1. Digital Literacy and Technology Skills:
Proficiency in digital tools and technologies is fundamental in today's job market. Courses in coding, data analysis, cybersecurity, and other tech-related fields are crucial for staying relevant.

2. Soft Skills:
Skills such as communication, teamwork, leadership, and emotional intelligence are highly valued across all industries. Developing these skills enhances one's ability to collaborate, manage, and lead effectively.

3. Industry-Specific Knowledge:
Keeping abreast of developments and innovations within one's industry is essential. Professional certifications, workshops, and seminars can provide deep insights into

specialized fields.

4. Adaptability and Learning Agility: The ability to learn quickly and adapt to new situations is a crucial skill in itself. Lifelong learners are often more adept at picking up new technologies, processes, and methodologies.

Strategies for Implementing Lifelong Learning and Reskilling

1. Formal Education and Training: Enrolling in degree programs, vocational training, and certification courses provides structured learning opportunities. Many institutions offer flexible learning options, such as online courses, to accommodate working professionals.

2. On-the-Job Training: Many companies provide on-the-job training programs to help employees develop new skills relevant to their roles. These programs can include mentorship, cross-training, and hands-on projects.

3. Self-Directed Learning: Individuals can take charge of their own learning through online resources, books, webinars, and tutorials. Platforms like Coursera, edX, and LinkedIn Learning offer a wide range of courses across various subjects.

4. Professional Development Programs:

Participating in workshops, conferences, and industry events provides opportunities for networking and learning from experts in the field. Professional associations often offer resources and programs aimed at career development.

5. Government and Organizational Support:

Governments and organizations can support lifelong learning and reskilling through policies and initiatives that provide funding, incentives, and infrastructure for education and training programs.

Lifelong learning and reskilling are essential for navigating the complexities of the modern job market and ensuring long-term career success. By committing to continuous skill development, individuals can adapt to technological advancements, economic shifts, and evolving industry demands. Strategies such as formal education, on-the-job training, self-directed learning, professional development, and supportive policies create a robust framework for lifelong learning and reskilling. Embracing these practices not only enhances professional resilience and employability but also contributes to personal growth and fulfillment.

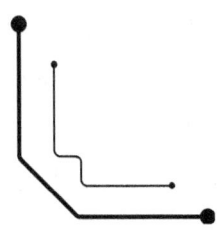

Chapter 5.
Opportunities and Challenges: The Dual Impact of AI

Job displacement vs. job creation

As technological advancements and automation continue to transform industries, the dynamics of job displacement and job creation become increasingly significant. Understanding the interplay between these two phenomena is essential for policymakers, businesses, and workers to navigate the changing employment landscape.

Job Displacement

Job displacement refers to the loss of jobs due to technological advancements, automation, and changes in industry practices. As new technologies and processes are introduced, certain roles become redundant or are significantly altered, leading to the displacement of workers who previously held those positions.

1. Automation and AI:
Automation and artificial intelligence (AI) are major drivers of job displacement. Tasks that were once performed by humans are increasingly being handled by machines and software, especially those involving repetitive, routine, and predictable activities. Examples include manufacturing assembly lines, data entry, and basic customer service functions.

2. Economic and Industry Shifts:

Changes in economic conditions and industry structures can also lead to job displacement. For instance, the decline of coal mining due to the shift towards renewable energy sources has resulted in the displacement of workers in the fossil fuel industry.

3. Skill Mismatches:

As industries evolve, there may be a mismatch between the skills workers possess and the skills required for new or changing roles. Workers who cannot adapt or reskill may face displacement.

Job Creation

Job creation refers to the generation of new jobs as a result of technological innovation, economic growth, and the emergence of new industries. While automation and AI can displace certain jobs, they also create opportunities in other areas.

1. New Industries and Roles:

Technological advancements often give rise to entirely new industries and roles. For example, the growth of the tech industry has led to the creation of jobs in software development, cybersecurity, data science, and digital marketing.

2. Increased Productivity and Demand:

Automation and AI can increase productivity and reduce

costs, leading to higher economic growth and increased demand for goods and services. This, in turn, can create new jobs in various sectors, including those related to the development, maintenance, and oversight of new technologies.

3. Enhanced Human-Machine Collaboration:
Many new roles involve working alongside advanced technologies rather than being replaced by them. Jobs in areas such as robotics maintenance, AI ethics, and machine learning operations exemplify how human skills are complemented by technological capabilities.

Balancing Job Displacement and Job Creation

1. Reskilling and Upskilling:
To mitigate job displacement, a focus on reskilling and upskilling the workforce is essential. Providing training programs and educational opportunities helps workers acquire the skills needed for emerging roles and industries.

2. Lifelong Learning:
Promoting a culture of lifelong learning ensures that workers continuously develop new skills throughout their careers. This adaptability is crucial in a rapidly changing job market.

3. Supportive Policies:
Governments and organizations play a key role in balancing job displacement and creation. Policies that

encourage innovation while providing safety nets for displaced workers, such as unemployment benefits, retraining programs, and job placement services, are vital.

4. Inclusive Economic Growth:
Ensuring that the benefits of technological advancements are widely distributed helps to create a more inclusive economy. This includes investing in infrastructure, education, and policies that support small and medium-sized enterprises (SMEs).

Case Studies and Examples

1. Manufacturing:
In manufacturing, automation has led to the displacement of many assembly line workers. However, it has also created new roles in robotics engineering, automation maintenance, and advanced manufacturing techniques.

2. Retail:
The rise of e-commerce has disrupted traditional retail jobs. Yet, it has also generated new opportunities in logistics, warehouse management, online customer service, and digital marketing.

3. Healthcare:
AI and automation are transforming healthcare by automating diagnostics and administrative tasks. This shift is creating demand for new roles in health informatics, telemedicine, and AI-assisted medical research.

The relationship between job displacement and job creation is complex and multifaceted. While technological advancements and economic shifts can lead to the loss of certain jobs, they also create new opportunities and roles. To navigate this evolving landscape, it is crucial to invest in reskilling and upskilling the workforce, promote lifelong learning, implement supportive policies, and strive for inclusive economic growth. By understanding and addressing the challenges and opportunities associated with job displacement and creation, society can ensure a more resilient and adaptable workforce for the future.

The gig economy and new employment models

The gig economy and new employment models are reshaping the landscape of work in significant ways. These emerging trends offer both opportunities and challenges for workers, employers, and policymakers. Understanding the dynamics of the gig economy and the implications of new employment models is essential for navigating this evolving workforce environment.

The Gig Economy

1. Definition and Characteristics:
The gig economy refers to a labor market characterized by short-term contracts or freelance work as opposed to permanent jobs. Gig workers typically engage in

temporary, flexible jobs, often facilitated by digital platforms like Uber, Lyft, Upwork, and TaskRabbit.

2. Flexibility and Autonomy:
One of the main attractions of the gig economy is the flexibility it offers. Workers can choose when, where, and how much they work, allowing them to balance personal and professional commitments more effectively. This autonomy appeals particularly to individuals seeking greater control over their work-life balance.

3. Diverse Opportunities:
The gig economy encompasses a wide range of activities, from driving and delivery services to freelance writing, graphic design, consulting, and skilled trades. This diversity allows individuals to leverage their unique skills and interests in various ways.

4. Income Variability:
While the gig economy offers flexibility, it often comes with income instability. Gig workers may face unpredictable earnings, lack of benefits such as health insurance and retirement plans, and less job security compared to traditional employment.

New Employment Models

1. Remote Work:
The COVID-19 pandemic accelerated the adoption of remote work, making it a permanent fixture in many

industries. Remote work models allow employees to work from anywhere, leveraging digital tools and communication platforms to stay connected and productive.

2. Hybrid Work:

Hybrid work models combine elements of remote work and in-office work. Employees may split their time between working from home and the office, providing a balance of flexibility and face-to-face interaction. This model aims to enhance productivity, employee satisfaction, and work-life balance.

3. Project-Based Employment:

Project-based employment involves hiring individuals or teams to complete specific projects rather than filling permanent positions. This model is common in industries like construction, IT, and creative fields, where work is often organized around discrete projects with defined timelines and deliverables.

4. Part-Time and Temporary Work:

Part-time and temporary employment models provide opportunities for workers who do not seek full-time commitments. These models can accommodate seasonal demands, project-specific needs, and other temporary requirements.

Implications for Workers

1. Skill Development and Adaptability:
Workers in the gig economy and new employment models must continuously develop and adapt their skills to stay competitive. This often involves self-directed learning and professional development.

2. Financial Management:
Gig workers and those in non-traditional employment models need to be adept at managing irregular income streams. Budgeting, saving, and understanding tax obligations are critical for financial stability.

3. Benefits and Protections:
A significant challenge for gig workers is the lack of traditional employment benefits such as health insurance, retirement plans, and paid leave. Policymakers and companies are exploring ways to provide these protections to non-traditional workers.

Implications for Employers

1. Access to Talent:
Employers benefit from the gig economy and new employment models by gaining access to a larger and more diverse talent pool. This can be particularly advantageous for specialized skills and short-term projects.

2. Cost Efficiency:
Hiring gig workers or adopting project-based employment

can be more cost-effective than maintaining a large permanent workforce. Employers can scale their workforce up or down based on demand, reducing overhead costs.

3. Workforce Management:
Managing a workforce that includes gig workers, remote employees, and project-based teams requires new strategies and tools. Effective communication, collaboration platforms, and performance management systems are essential for success.

Policy and Regulatory Considerations

1. Worker Protections:
Policymakers are increasingly focused on ensuring that gig workers and those in new employment models receive adequate protections and benefits. This includes discussions around minimum wage laws, health benefits, and retirement savings options.

2. Labor Laws:
Existing labor laws may need to be updated to reflect the realities of the gig economy and new employment models. This includes redefining employment classifications and ensuring fair labor practices.

3. Social Safety Nets:
Strengthening social safety nets to support gig workers and those in non-traditional employment is crucial. Unemployment benefits, healthcare access, and income

support programs need to be inclusive of all types of workers.

The gig economy and new employment models are transforming the way work is organized and performed. While these trends offer flexibility, autonomy, and access to diverse opportunities, they also present challenges related to income stability, benefits, and worker protections. For workers, continuous skill development and financial management are key to navigating this landscape. For employers, effectively managing a diverse and dynamic workforce is essential. Policymakers play a critical role in ensuring that labor laws and social safety nets evolve to support the needs of all workers in this new employment paradigm. By understanding and addressing these implications, society can harness the benefits of the gig economy and new employment models while mitigating their challenges.

Addressing inequality and access to opportunities

Addressing inequality and ensuring access to opportunities are critical goals for fostering a fair and inclusive society. As technological advancements and economic shifts reshape the global landscape, it is essential to implement strategies that promote equity and enable all individuals to thrive. Here are key areas and strategies to consider:

Understanding Inequality

1. Economic Disparities:
Economic inequality manifests in the unequal distribution of wealth and income. Factors such as educational background, employment opportunities, and access to resources contribute to these disparities.

2. Social Inequality:
Social inequality includes differences in status, power, and access to social services. This can be influenced by factors such as race, gender, ethnicity, and social class.

3. Digital Divide:
The digital divide refers to the gap between individuals who have access to modern information and communication technology and those who do not. This divide exacerbates existing inequalities, limiting opportunities for education, employment, and social engagement.

Promoting Equal Access to Education

1. Quality Education for All:
Ensuring that all individuals have access to quality education is foundational for addressing inequality. This includes improving schools in underserved areas, providing adequate resources, and ensuring well-trained teachers.

2. Affordable Higher Education:

Making higher education affordable and accessible is crucial. Scholarships, grants, and low-interest student loans can help reduce financial barriers for students from low-income families.

3. Lifelong Learning Opportunities:

Providing opportunities for lifelong learning helps individuals continuously update their skills and knowledge, which is essential in a rapidly changing job market. This includes adult education programs, vocational training, and online courses.

Enhancing Employment Opportunities

1. Job Creation:

Policies that stimulate job creation, especially in underserved areas, can reduce unemployment and underemployment. This includes investing in infrastructure, supporting small businesses, and promoting entrepreneurship.

2. Fair Wages and Benefits:

Ensuring fair wages and benefits for all workers is crucial for reducing economic inequality. This includes advocating for a living wage, health benefits, and paid leave.

3. Inclusive Hiring Practices:

Encouraging inclusive hiring practices helps ensure that all individuals, regardless of their background, have access to employment opportunities. This includes addressing biases in recruitment and promoting diversity in the workplace.

Bridging the Digital Divide

1. Access to Technology:
Providing affordable access to technology, such as computers and high-speed internet, is essential. Programs that subsidize technology costs for low-income households can help bridge the digital divide.

2. Digital Literacy Training:
Offering digital literacy training ensures that individuals can effectively use technology for education, employment, and personal development. This includes training in basic computer skills, internet navigation, and cybersecurity.

3. Community Tech Centers:
Establishing community technology centers provides access to digital tools and training in underserved areas. These centers can serve as hubs for learning, job searching, and social interaction.

Supporting Social and Economic Mobility

1. Affordable Housing:
Ensuring access to affordable housing helps reduce economic disparities and supports social mobility. Policies that promote affordable housing development and provide housing assistance to low-income families are

essential.

2. Access to Healthcare:
Providing affordable and accessible healthcare reduces social inequalities and supports overall well-being. This includes expanding health insurance coverage, offering preventive care, and addressing social determinants of health.

3. Transportation Access:
Improving transportation infrastructure and ensuring affordable public transit options can enhance access to employment, education, and other essential services.

Policy and Government Intervention

1. Progressive Taxation:
Implementing progressive taxation policies ensures that wealthier individuals contribute a fair share to funding public services and social programs that benefit all citizens.

2. Social Safety Nets:
Strengthening social safety nets, such as unemployment benefits, food assistance, and child care support, helps protect individuals from economic hardships and promotes social equity.

3. Anti-Discrimination Laws:
Enforcing anti-discrimination laws ensures that all

individuals are treated fairly and have equal access to opportunities, regardless of their race, gender, or socioeconomic background.

Community and Grassroots Initiatives

1. Local Advocacy:
Supporting local advocacy groups and community organizations that work to address inequality and promote access to opportunities is crucial. These groups often have a deep understanding of local needs and can drive effective change.

2. Public Awareness Campaigns:
Raising awareness about the importance of equality and access to opportunities can galvanize public support and encourage action. Public awareness campaigns can highlight success stories, educate about existing disparities, and promote inclusive policies.

3. Collaborative Efforts:
Encouraging collaboration between government, private sector, and non-profit organizations can lead to more comprehensive and effective solutions. Public-private partnerships can leverage resources and expertise to address complex issues related to inequality.

Addressing inequality and ensuring access to opportunities require a multifaceted approach involving education, employment, technology, healthcare, housing, and policy interventions. By implementing strategies that promote equity and inclusion, society can create an environment where all individuals have the chance to succeed and contribute to their communities. Ensuring that these efforts are sustained and supported by collaborative efforts among various stakeholders is essential for long-term impact.

Chapter 6.
Policy and Regulation: Governing the AI Era

Government roles and responsibilities

Governments play a crucial role in shaping society and ensuring the well-being of their citizens. Their responsibilities span a wide range of areas, from maintaining public order and providing public services to ensuring economic stability and protecting individual rights. Here's an overview of the key roles and responsibilities of government:

Maintaining Public Order and Safety

1. Law Enforcement:
Governments are responsible for creating and enforcing laws to maintain public order. This includes policing, criminal justice systems, and correctional facilities to ensure the rule of law is upheld and citizens are protected from crime.

2. Emergency Services:
Providing emergency services such as fire departments, paramedics, and disaster response teams is essential. These services ensure rapid response to emergencies, safeguarding lives and property.

Providing Public Services

1. Healthcare:
Governments often oversee and fund public healthcare systems to ensure that all citizens have access to medical care. This includes hospitals, clinics, vaccination programs, and public health initiatives.

2. Education:
Ensuring access to quality education is a fundamental government responsibility. This includes funding and regulating public schools, colleges, and universities, and providing scholarships and grants to support students.

3. Infrastructure:
Governments are responsible for building and maintaining essential infrastructure such as roads, bridges, public transportation, water supply, and sewage systems. This infrastructure is vital for economic development and public well-being.

Economic Management

1. Fiscal Policy:
Governments use fiscal policy to manage the economy by adjusting spending levels and tax rates. This includes creating budgets, collecting taxes, and allocating funds for public projects and services.

2. Monetary Policy:

While often managed by independent central banks, monetary policy is also a critical aspect of economic management. This includes regulating the money supply and interest rates to control inflation and stabilize the economy.

3. Regulation and Oversight:

Governments regulate industries and financial markets to ensure fair practices, competition, and consumer protection. This includes enforcing labor laws, environmental regulations, and trade policies.

Social Welfare and Protection

1. Social Security:

Providing social security systems, such as pensions, unemployment benefits, and disability benefits, helps protect citizens from economic hardship.

2. Public Assistance Programs:

Governments implement public assistance programs, such as food assistance, housing subsidies, and healthcare subsidies, to support low-income individuals and families.

3. Child and Family Services:

Protecting vulnerable populations, such as children, the elderly, and those with disabilities, is a key responsibility. This includes child welfare services, elder care programs, and support for disabled individuals.

Environmental Protection and Sustainability

1. Conservation and Preservation:
Governments play a crucial role in conserving natural resources and protecting ecosystems. This includes managing national parks, wildlife reserves, and marine sanctuaries.

2. Environmental Regulations:
Implementing and enforcing environmental regulations to reduce pollution, protect natural habitats, and combat climate change is essential for sustainable development.

3. Sustainable Development:
Promoting sustainable development practices ensures that economic growth does not come at the expense of environmental health. This includes investing in renewable energy, sustainable agriculture, and green technologies.

Protecting Individual Rights and Freedoms

1. Civil Rights:
Governments are responsible for protecting civil rights and ensuring equality under the law. This includes enforcing anti-discrimination laws, voting rights, and freedom of speech.

2. Justice System:
Providing a fair and impartial justice system is essential for upholding the rule of law. This includes courts, legal aid,

and public defenders to ensure access to justice for all citizens.

3. Privacy and Security:
Balancing national security needs with the protection of individual privacy and civil liberties is a critical government responsibility. This includes regulations on data protection and surveillance.

International Relations and Diplomacy

1. Foreign Policy:
Governments conduct foreign policy to manage relationships with other countries. This includes diplomacy, trade negotiations, and participation in international organizations.

2. National Defense:
Ensuring national security through defense forces and intelligence services is a fundamental government responsibility. This includes maintaining military readiness and protecting against external threats.

3. International Aid and Development:
Providing international aid and supporting global development initiatives helps foster international cooperation and address global challenges such as poverty, disease, and climate change.

The roles and responsibilities of government are vast and

multifaceted, encompassing the maintenance of public order, provision of essential services, economic management, social welfare, environmental protection, safeguarding individual rights, and conducting international relations. Effective governance requires balancing these diverse responsibilities to promote the well-being of citizens, ensure sustainable development, and maintain a stable and just society. By fulfilling these roles, governments can create a foundation for prosperity, security, and equality.

Regulatory frameworks and ethical guidelines

Regulatory frameworks and ethical guidelines are essential components in the governance of industries, technologies, and societal practices. They ensure that activities are conducted in a manner that is safe, fair, and respectful of ethical standards. Understanding their roles and the interplay between regulation and ethics is crucial for fostering responsible and sustainable practices.

Regulatory Frameworks

1. Definition and Purpose:
Regulatory frameworks consist of laws, regulations, and rules established by governments or regulatory bodies to control and guide the activities within various industries. Their primary purposes are to ensure safety, protect

consumers, maintain fair competition, and promote transparency.

2. Implementation and Enforcement:
Regulatory agencies are responsible for the implementation and enforcement of these frameworks. This includes monitoring compliance, conducting inspections, and imposing penalties for violations. Examples include the Food and Drug Administration (FDA) in the United States, which regulates food and pharmaceuticals, and the Environmental Protection Agency (EPA), which oversees environmental protection.

3. Industry-Specific Regulations:
Different industries require tailored regulatory frameworks to address their unique challenges and risks. For instance:

-**Healthcare:** Regulations ensure the safety and efficacy of medical treatments and protect patient privacy through laws like the Health Insurance Portability and Accountability Act (HIPAA).

-**Finance:** Financial regulations aim to maintain market integrity, protect investors, and prevent fraud. Examples include the Dodd-Frank Act and regulations by the Securities and Exchange Commission (SEC).

-**Technology:** The rapidly evolving tech industry is regulated to protect data privacy, ensure cybersecurity, and manage the ethical use of artificial intelligence (AI). Regulations like the General Data Protection Regulation (GDPR) in the EU set standards for data protection.

Ethical Guidelines

1. Definition and Importance:
Ethical guidelines are principles and standards that guide the conduct of individuals and organizations, ensuring that their actions are morally sound and socially responsible. They are crucial for maintaining trust and integrity within industries and society.

2. Development of Ethical Guidelines:
Ethical guidelines are often developed by professional bodies, industry associations, and organizations. They may also be influenced by cultural values, philosophical principles, and societal norms. Examples include the American Medical Association's Code of Medical Ethics and the IEEE's guidelines for ethical AI.

3. Core Ethical Principles:

-**Autonomy:** Respecting the autonomy of individuals by ensuring they have the freedom to make informed decisions.

-**Beneficence:** Promoting the well-being of individuals and communities by acting in their best interests.
Non-Maleficence: Avoiding harm to individuals and

communities.

-**Justice:** Ensuring fairness and equality in the distribution of benefits and burdens.

Interplay Between Regulatory Frameworks and Ethical Guidelines

1. Complementary Roles:
Regulatory frameworks and ethical guidelines often complement each other. Regulations provide the legal backbone to ensure compliance and protect public interests, while ethical guidelines offer a moral compass for behavior that may not be explicitly covered by laws.

2. Navigating Gray Areas:
In areas where regulations may be lacking or slow to catch up with technological advancements, ethical guidelines help navigate gray areas and promote responsible conduct. For instance, in AI development, ethical principles guide practices in fairness, accountability, and transparency, even in the absence of comprehensive legislation.

3. Proactive and Reactive Measures:
Regulations are often reactive, responding to identified risks and issues. In contrast, ethical guidelines can be proactive, encouraging best practices and preventing harm before it occurs. Together, they create a robust framework for governance.

Challenges and Considerations

1. Balancing Innovation and Regulation:
One of the primary challenges is striking a balance between fostering innovation and implementing regulations that protect public interests. Over-regulation can stifle innovation, while under-regulation can lead to ethical lapses and harm.

2. Global Consistency:
With globalization, ensuring consistent regulatory and ethical standards across different countries is challenging. International cooperation and harmonization of standards are essential for managing cross-border activities and ensuring global accountability.

3. Adapting to Change:
Rapid technological advancements require regulatory frameworks and ethical guidelines to be adaptable and forward-thinking. Continuous review and updating of these frameworks are necessary to keep pace with new developments and emerging risks.

Regulatory frameworks and ethical guidelines are integral to the responsible governance of industries and technologies. They provide the structure and principles necessary to ensure that activities are conducted safely, fairly, and ethically. By working together, regulations and ethics create a comprehensive system that promotes trust, accountability, and sustainable progress. Balancing these elements, addressing global consistency, and adapting to change are crucial for navigating the complexities of the modern world and ensuring the well-being of society.

Chapter 7.
Remote Work and Flexibility: The New Norm?

Rise of remote work due to technology

The rise of remote work is one of the most significant transformations in the modern workplace, driven largely by advancements in technology. This shift has reshaped how businesses operate, how employees interact, and how work itself is structured. Here's an in-depth look at the factors behind this change, the benefits and challenges it presents, and its future implications.

Technological Drivers of Remote Work

1. Internet Connectivity:
High-speed internet has become widely accessible, enabling employees to work from virtually anywhere. Reliable internet connections allow for seamless communication, access to cloud-based tools, and efficient collaboration.

2. Cloud Computing:
Cloud services like Google Workspace, Microsoft 365, and AWS enable employees to access documents, software, and applications from any location. This flexibility supports remote work by allowing teams to share and edit documents in real-time, regardless of their physical location.

3. Communication Tools:

Platforms such as Zoom, Microsoft Teams, and Slack have revolutionized communication. Video conferencing, instant messaging, and virtual meeting tools facilitate real-time interaction and collaboration, making remote work feasible and productive.

4. Project Management Software:

Tools like Asana, Trello, and Jira help teams manage projects, track progress, and meet deadlines without being in the same physical space. These platforms provide visibility into tasks and workflows, ensuring that everyone stays aligned and accountable.

5. Cybersecurity Advancements:

Enhanced cybersecurity measures protect sensitive data and ensure secure access to company resources. Virtual private networks (VPNs), encryption, and multi-factor authentication are critical for maintaining data integrity and security in a remote work environment.

Benefits of Remote Work

1. Increased Flexibility:

Remote work offers employees the flexibility to create their own schedules and work from locations that suit their personal and professional needs. This flexibility can lead to improved work-life balance and higher job satisfaction.

2. Access to a Broader Talent Pool:

Employers can recruit talent from around the globe, not limited by geographical boundaries. This expands the talent pool, allowing companies to find the best candidates regardless of their location.

3. Cost Savings:

Both employers and employees can save on costs. Companies can reduce expenses related to office space, utilities, and on-site amenities, while employees can save on commuting, work attire, and meals.

4. Increased Productivity:

Many employees report higher productivity levels when working remotely, due to fewer office distractions and the ability to create a personalized work environment.

5. Environmental Impact:

Reduced commuting leads to lower carbon emissions, contributing to environmental sustainability. Remote work can play a part in reducing the overall carbon footprint of businesses.

Challenges of Remote Work

1. Communication Barriers:

Despite advanced communication tools, remote work can lead to misunderstandings and miscommunication. The lack of face-to-face interaction can make it harder to convey tone, context, and nuance.

2. Collaboration Difficulties:
Collaborative tasks that require spontaneous brainstorming or quick feedback can be more challenging in a remote setting. Teams may need to make extra efforts to maintain strong collaboration.

3. Isolation and Loneliness:
Remote work can lead to feelings of isolation and loneliness, as employees miss out on the social interactions and camaraderie of an office environment.

4. Work-Life Boundaries:
Without a clear separation between work and home, some remote workers may struggle to maintain boundaries, leading to overwork and burnout.

5. Technical Issues:
Reliable technology is crucial for remote work, but not all employees have access to high-quality equipment or stable internet connections. Technical issues can disrupt productivity and create frustration.

Future Implications of Remote Work

1. Hybrid Work Models:
The future of work is likely to include hybrid models, combining remote and on-site work. This approach allows for flexibility while maintaining opportunities for in-person collaboration and team building.

2. Redefining Office Spaces:
Traditional office spaces may be reimagined as collaborative hubs rather than primary work locations. Companies might invest in co-working spaces or smaller, satellite offices to accommodate a distributed workforce.

3. Evolving Management Practices:
Managers will need to adapt their leadership styles to effectively manage remote teams. This includes focusing on outcomes rather than hours worked, fostering a strong virtual team culture, and providing support for remote work challenges.

4. Policy and Regulation Changes:
Governments and organizations will need to update policies and regulations to address issues such as remote work rights, data security, taxation, and employee benefits.

5. Technological Innovation:
Continued advancements in technology will further enhance remote work capabilities. Innovations in virtual reality (VR), augmented reality (AR), and artificial intelligence (AI) could create new ways for remote teams to collaborate and connect.

The rise of remote work due to technological advancements represents a profound shift in how we approach work. While it offers numerous benefits such as increased flexibility, access to a global talent pool, and cost savings, it also presents challenges like communication barriers and potential isolation. As remote work continues to evolve, businesses, employees, and policymakers must collaborate to develop effective strategies and solutions that leverage technology while addressing the human aspects of work. By doing so, we can create a more adaptable, inclusive, and productive future of work.

Benefits and Drawbacks of Remote and Flexible Work

Remote and flexible work arrangements have transformed the traditional workplace landscape, offering a range of benefits and drawbacks for both employers and employees. Understanding these pros and cons can help organizations and workers make informed decisions about adopting and managing remote and flexible work models.

Benefits of Remote and Flexible Work

1. Increased Flexibility and Work-Life Balance:

-**Autonomy**: Employees can often set their own schedules, which helps them manage personal commitments alongside professional responsibilities.

-**Work-Life Integration**: The ability to work from home or other locations allows for better integration of work and personal life, reducing stress and increasing overall life satisfaction.

2. Enhanced Productivity and Performance:

-**Fewer Distractions:** Many remote workers find that they can focus better without the interruptions common in office environments.

-**Personalized Work Environment:** Employees can create a work setup that suits their individual needs and preferences, potentially leading to increased productivity.

3. Cost Savings:

-**Reduced Commuting Costs:** Employees save money on transportation, meals, and work attire.

-**Lower Overhead for Employers:** Companies can save on office space, utilities, and other overhead costs associated with maintaining a physical workplace.

4. Access to a Broader Talent Pool:

-**Geographic Diversity:** Employers can hire the best talent regardless of their location, increasing diversity and inclusion.

-**Retention and Attraction:** Offering flexible work arrangements can attract new talent and retain existing employees who value flexibility.

5. Environmental Benefits:

-Reduced Carbon Footprint: Fewer commutes result in lower greenhouse gas emissions and decreased environmental impact.

Drawbacks of Remote and Flexible Work

1. Communication and Collaboration Challenges:

-Lack of Face-to-Face Interaction: Remote work can hinder spontaneous communication and make it harder to build relationships and trust among team members.

-Coordination Issues: Time zone differences and varying schedules can complicate coordination and collaboration on projects.

2. Potential for Isolation and Loneliness:

-Social Isolation: Employees working remotely may miss out on social interactions and the camaraderie of an office environment, leading to feelings of loneliness.

-Mental Health Concerns: The lack of social support and human connection can negatively impact mental health and well-being.

3. Work-Life Boundaries:

-**Blurring of Boundaries:** Without clear separation between work and home life, employees may struggle to "switch off," leading to overwork and burnout.

-**Family Distractions:** Working from home can introduce distractions from family members or household responsibilities.

4. Dependence on Technology:

-**Technical Issues:** Reliable technology and internet access are crucial for remote work. Technical problems can disrupt productivity and create frustration.

-**Cybersecurity Risks:** Remote work increases the risk of data breaches and cyberattacks, requiring robust cybersecurity measures.

5. Management and Performance Monitoring:

-**Challenges in Supervision:** Managers may find it difficult to oversee and support remote workers effectively, especially in ensuring accountability and maintaining team cohesion.

-**Performance Measurement:** Traditional performance metrics may not apply, necessitating new approaches to evaluate employee performance and productivity.

Strategies for Mitigating Drawbacks

1. Effective Communication Tools: Utilize advanced communication platforms like Slack, Zoom, and Microsoft Teams to facilitate clear and consistent communication.

2. Regular Check-Ins: Schedule regular meetings and check-ins to maintain connection, provide support, and address any issues promptly.

3. Clear Policies and Expectations: Establish clear guidelines for remote work, including expectations around availability, communication, and performance metrics.

4. Support for Mental Health: Provide resources and support for mental health, such as access to counseling services, wellness programs, and promoting a healthy work-life balance.

5. Enhanced Cybersecurity: Implement robust cybersecurity measures, including VPNs, secure login protocols, and regular security training for employees.

Remote and flexible work offer numerous benefits, including increased flexibility, cost savings, and access to a broader talent pool. However, these advantages come with challenges such as communication barriers, potential isolation, and cybersecurity risks. By understanding and addressing these drawbacks, organizations can create effective remote work policies that maximize the benefits while minimizing the downsides. Adopting a thoughtful and strategic approach to remote and flexible work can lead to a more satisfied, productive, and resilient workforce.

Future projections for workplace flexibility

The future of workplace flexibility promises significant transformations driven by technological advancements, shifting cultural attitudes, and evolving business needs. Here are some key projections for how workplace flexibility is likely to evolve in the coming years:

1. Increased Adoption of Hybrid Work Models

Hybrid work models, which combine remote and in-office work, are expected to become the norm. Companies will offer employees the flexibility to choose where they work based on their tasks, preferences, and collaboration needs.

-**Personalized Work Schedules:** Employees will have more control over their work hours, allowing them to balance personal and professional responsibilities effectively.

-**Office Spaces as Collaboration Hubs:** Physical office spaces will be reimagined as places for team collaboration, brainstorming sessions, and social interactions, rather than daily workstations.

2. Technological Advancements

Technology will continue to play a pivotal role in enabling and enhancing workplace flexibility. Several technological trends are likely to shape the future of work:

-**Advanced Collaboration Tools:** Innovations in collaboration software, such as virtual reality (VR) and augmented reality (AR), will create immersive meeting experiences, making remote interactions more engaging and effective.

-**AI and Automation:** Artificial intelligence (AI) and automation will handle routine tasks, allowing employees to focus on more complex and creative work. AI-driven analytics will also help in monitoring productivity and well-being.

-**Enhanced Cybersecurity:** As remote work persists, robust cybersecurity measures will be essential to protect sensitive data and maintain trust.

3. Evolution of Management Practices

Effective management will require new strategies to support flexible work arrangements. Managers will need to adapt their approaches to foster a productive and cohesive remote workforce.

-**Outcome-Based Performance Measurement:** Emphasis will shift from hours worked to outcomes achieved, with clear goals and key performance indicators (KPIs) guiding employee evaluations.

-**Virtual Leadership Skills:** Managers will need to develop skills in virtual communication, remote team-building, and providing support in a decentralized work environment.

-**Employee Well-Being Focus:** There will be greater emphasis on mental health and well-being, with managers playing a crucial role in recognizing and addressing signs of burnout or isolation.

4. Changes in Organizational Culture

Workplace flexibility will necessitate cultural shifts within organizations to ensure inclusivity, collaboration, and mutual trust.

-**Inclusivity and Diversity**: Flexible work arrangements can promote diversity by enabling people from different geographic locations and backgrounds to join the workforce.

-**Trust and Autonomy:** A culture of trust and autonomy will be critical, as employees will need to manage their own schedules and productivity.

-**Continuous Learning and Development:** Lifelong learning will become a cornerstone of organizational culture, with companies investing in training programs to keep employees' skills relevant in a rapidly changing work environment.

5. Policy and Regulatory Developments

Governments and regulatory bodies will need to adapt to the changing nature of work, creating policies that support and protect flexible work arrangements.

-**Remote Work Legislation:** New laws and regulations will address remote work rights, data protection, and employer responsibilities to ensure fair treatment and security.

-**Tax and Benefits Considerations:** Policies will need to consider taxation for remote workers and ensure they have access to benefits such as health insurance, retirement plans, and paid leave.

-**Support for Small Businesses:** Small and medium-sized enterprises (SMEs) will need guidance and support to implement flexible work models effectively.

6. Impact on Urban and Rural Development

The widespread adoption of flexible work will influence urban and rural development patterns.

-**Decentralization of Cities:** As more people work remotely, there may be a shift away from densely populated urban centers, leading to reduced congestion and demand for office space in cities.

-**Revitalization of Rural Areas:** Rural and suburban areas could see growth as people seek more affordable living conditions and a better quality of life,

Workplace flexibility is poised to become a defining feature of the future of work. By embracing hybrid work models, leveraging technological advancements, evolving management practices, fostering inclusive cultures, and adapting policies, organizations can create flexible work environments that benefit both employers and employees. These changes will not only enhance productivity and job satisfaction but also contribute to a more resilient and adaptive workforce ready to meet the challenges of the future.

Chapter 8.
Mental Health and Well-being: Navigating Change

Impact of automation on mental health

The integration of automation into the workplace brings significant changes to how tasks are performed, influencing various aspects of employee mental health. While automation offers numerous benefits such as increased efficiency and reduced manual labor, it also presents challenges that can impact mental well-being. Understanding these impacts is crucial for developing strategies to support employees in an increasingly automated work environment.

Positive Impacts on Mental Health

1. Reduction in Repetitive Tasks:
Automation can alleviate the burden of monotonous, repetitive tasks, allowing employees to engage in more meaningful and intellectually stimulating work. This shift can enhance job satisfaction and reduce feelings of boredom and frustration.

2. Improved Work-Life Balance:
By automating routine processes, employees may

experience reduced workloads and more flexible schedules. This can lead to improved work-life balance, giving employees more time for personal activities, relaxation, and family, which are vital for mental health.

3. Enhanced Productivity and Efficiency:
Automation can boost overall productivity, leading to a sense of accomplishment and fulfillment. When employees see the positive outcomes of their work facilitated by automation, it can enhance their morale and motivation.

4. Reduction in Work-Related Stress:
Automation can help minimize errors and increase accuracy in tasks, reducing the stress and anxiety associated with making mistakes. This reliability can create a more stable and predictable work environment, contributing to lower stress levels.

Negative Impacts on Mental Health

1. Job Insecurity and Anxiety:
The fear of job displacement due to automation is a significant concern for many employees. Worries about redundancy and the potential for job loss can lead to chronic anxiety, stress, and a sense of insecurity about the future.

2. Skill Obsolescence and Pressure to Upskill:
Rapid technological advancements necessitate continuous learning and adaptation. Employees may feel pressured to

constantly update their skills to remain relevant, leading to stress and anxiety about keeping pace with automation trends.

3. Isolation and Reduced Social Interaction:
Increased reliance on automated systems can lead to reduced human interaction in the workplace. Social connections and interactions are important for mental health, and their reduction can result in feelings of isolation and loneliness.

4. Over-Reliance on Technology:
Dependence on automated systems can create stress when these systems fail or encounter issues. Employees may experience frustration and anxiety when technological problems disrupt their workflow and productivity.

Strategies to Mitigate Negative Impacts

1. Clear Communication and Reassurance:
Employers should communicate transparently about the role of automation and its impact on job security. Reassuring employees about their value and the organization's commitment to their growth can alleviate anxiety.

2. Investment in Training and Development:
Providing opportunities for continuous learning and upskilling can empower employees to adapt to

technological changes confidently. Offering training programs and resources helps employees feel more competent and less threatened by automation.

3. Promoting Social Interaction:
Encouraging team-building activities, social events, and collaborative projects can maintain a sense of community and connection, even in an automated work environment. Hybrid work models can also help balance remote work with in-person interactions.

4. Supportive Mental Health Resources:
Employers should provide access to mental health resources, such as counseling services, wellness programs, and stress management workshops. Creating an open and supportive culture around mental health is crucial for addressing the challenges posed by automation.

5. Balanced Approach to Automation:
Implementing automation in a way that complements human work rather than replacing it entirely can help maintain job satisfaction and engagement. Combining human creativity and problem-solving with automated efficiency can create a more balanced and fulfilling work environment.

The impact of automation on mental health is multifaceted, with both positive and negative aspects. While automation can enhance job satisfaction, productivity, and work-life balance, it can also lead to anxiety, job insecurity, and isolation. By proactively addressing these challenges through clear communication, continuous learning opportunities, social interaction, and mental health support, organizations can create a healthier and more resilient workforce. Embracing a balanced approach to automation will help ensure that its benefits are maximized while mitigating its potential drawbacks on employee mental health.

Strategies for Maintaining Well-Being

Maintaining well-being, especially in the context of the rapidly evolving workplace and increased automation, requires a multifaceted approach that addresses both physical and mental health. Here are some key strategies for individuals and organizations to foster a healthy and balanced lifestyle:

1. Establishing a Healthy Work-Life Balance

-**Set Boundaries:** Clearly delineate work hours from personal time. Avoid checking emails or working on tasks outside of designated work hours to prevent burnout.

-**Take Breaks:** Incorporate regular breaks into your daily schedule. Short breaks can help refresh your mind and maintain productivity throughout the day.

- **Flexible Scheduling:** Leverage flexible work arrangements to manage personal responsibilities and reduce stress.

2. Promoting Physical Health

- **Regular Exercise:** Engage in regular physical activity, such as walking, jogging, or yoga, to boost physical and mental health. Exercise helps reduce stress, improve mood, and increase energy levels.

- **Healthy Eating:** Maintain a balanced diet rich in fruits, vegetables, lean proteins, and whole grains. Proper nutrition supports overall well-being and helps maintain energy levels.

- **Adequate Sleep:** Ensure you get enough quality sleep each night. Good sleep hygiene practices, such as a consistent sleep schedule and a relaxing bedtime routine, are essential for mental and physical health.

3. Fostering Mental Health

- **Mindfulness and Meditation:** Practice mindfulness techniques, such as meditation or deep-breathing exercises, to manage stress and enhance emotional regulation.

- **Limit Screen Time:** Reduce screen time, particularly before bedtime, to prevent eye strain and promote better sleep. Taking regular breaks from screens can also help prevent digital fatigue.

-Seek Support: Don't hesitate to seek professional help if you're feeling overwhelmed. Counseling or therapy can provide valuable tools and strategies for managing mental health challenges.

4. Encouraging Social Connections

-Build Relationships: Foster positive relationships with colleagues, friends, and family. Social support is crucial for emotional well-being.

-Participate in Team Activities: Engage in team-building activities and social events organized by your workplace to strengthen bonds with colleagues.

-Stay Connected: Use technology to stay connected with loved ones, especially in remote work settings, to combat feelings of isolation.

5. Continuous Learning and Personal Growth

-Lifelong Learning: Continuously seek opportunities for professional and personal development. Enroll in courses, attend workshops, or pursue new hobbies to keep your mind engaged and motivated.

-**Set Goals:** Establish short-term and long-term goals to provide direction and purpose. Achieving these goals can boost self-esteem and personal satisfaction.

6. Creating a Positive Work Environment

-**Ergonomic Workspace:** Set up an ergonomic workspace that promotes good posture and comfort, reducing the risk of physical strain.

-**Positive Work Culture:** Advocate for a positive work culture that values respect, inclusivity, and support. A healthy work environment can significantly impact mental well-being.

-**Recognition and Reward:** Regularly acknowledge and reward achievements and contributions. Feeling valued and appreciated enhances job satisfaction and motivation.

7. Stress Management Techniques

-**Time Management:** Utilize time management tools and techniques to organize tasks and prioritize work efficiently. Effective time management can reduce the feeling of being overwhelmed.

-**Relaxation Techniques:** Practice relaxation techniques such as progressive muscle relaxation, guided imagery, or listening to calming music to reduce stress levels.

-**Workload Management:** Communicate with supervisors and team members about workload expectations and seek support if you're feeling overburdened.

8. Leveraging Technology for Well-Being

-**Health Apps:** Use health and wellness apps to track physical activity, monitor sleep patterns, and provide guided meditation or relaxation exercises.

-**Online Resources:** Access online resources for mental health support, such as virtual counseling services or well-being webinars.

-**Automation Benefits:** Utilize automation to reduce the burden of repetitive tasks, allowing more time for creative and fulfilling activities.

Maintaining well-being in the modern workplace requires a comprehensive approach that integrates physical health, mental health, social connections, continuous learning, and effective stress management. By implementing these strategies, individuals can enhance their overall well-being and organizations can foster a supportive and healthy work environment. Embracing a holistic approach to well-being will help navigate the challenges of the evolving work landscape and promote a balanced, fulfilling life.

Organizational Responsibility and Support Systems

In the era of rapid technological change and evolving work environments, organizations have a crucial role in ensuring the well-being and development of their employees. By embracing their responsibilities and establishing comprehensive support systems, organizations can foster a healthy, productive, and resilient workforce. Here are key areas where organizational responsibility and support systems come into play:

1. Creating a Positive Work Culture

-Inclusive Environment: Cultivate an inclusive work culture that values diversity and ensures all employees feel respected and valued. Promote policies and practices that support equity and inclusion.

-Recognition and Appreciation: Implement regular recognition programs to celebrate employee achievements and contributions. Appreciation can significantly boost morale and motivation.

-Open Communication: Encourage open and transparent communication between management and employees. Create channels for feedback and ensure that employees feel heard and valued.

2. Promoting Employee Well-Being

-Mental Health Support: Provide access to mental health resources such as counseling services, employee assistance programs (EAPs), and mental health workshops. Normalize discussions about mental health to reduce stigma.

-Work-Life Balance: Encourage practices that promote work-life balance, such as flexible working hours, remote work options, and adequate vacation time. Support employees in managing their personal and professional lives.

-Health and Wellness Programs: Offer wellness programs that include physical fitness activities, health screenings, and nutritional guidance. Promote initiatives that support a healthy lifestyle.

3. Providing Learning and Development Opportunities

-Continuous Learning: Invest in continuous learning opportunities for employees. Offer training programs, workshops, and access to online courses to help employees develop new skills and advance their careers.

-Career Development: Implement career development programs that provide clear pathways for growth and progression within the organization. Offer mentorship and coaching to support employees in their professional journeys.

-Skill Reskilling and Upskilling: Focus on reskilling and upskilling initiatives to ensure employees remain competitive and adaptable in a rapidly changing work environment.

4. Ensuring Job Security and Fair Employment Practices

-Transparent Policies: Develop and communicate clear policies regarding job security, performance evaluations, and career progression. Transparency helps reduce anxiety and build trust.

-Fair Compensation: Ensure that compensation and benefits are fair, competitive, and aligned with industry standards. Regularly review and adjust pay scales to reflect the cost of living and market conditions.

-Job Stability: Provide job stability by anticipating changes brought by automation and implementing strategies to redeploy employees rather than resorting to layoffs.

5. Enhancing Workplace Safety and Ergonomics

-Safe Work Environment: Ensure a safe and healthy work environment by adhering to occupational safety and health regulations. Regularly assess and mitigate workplace hazards.

-**Ergonomic Workspaces:** Design ergonomic workspaces that reduce physical strain and promote employee comfort. Provide the necessary equipment and tools to support good posture and prevent injury.

-**Remote Work Safety:** Extend safety and ergonomic considerations to remote work environments by providing guidance and resources for setting up home offices.

6. Fostering Effective Leadership

-**Supportive Leadership:** Train leaders to be supportive, empathetic, and effective in managing teams. Good leadership is crucial for employee satisfaction and retention.

-**Conflict Resolution:** Equip managers with skills to handle conflicts effectively and fairly. Promote a culture where issues are addressed constructively and promptly.

-**Leadership Development:** Invest in leadership development programs to cultivate future leaders within the organization. Ensure that leaders are well-prepared to navigate the complexities of a modern workplace.

7. Leveraging Technology to Support Employees

-**Advanced Tools:** Provide employees with advanced

tools and technologies that enhance productivity and efficiency. Ensure that these tools are user-friendly and support seamless work processes.

-**Digital Well-Being:** Promote digital well-being by encouraging healthy technology use. Implement policies that prevent digital overload and support work-life balance.

-**Automation and AI:** Use automation and AI to support employees rather than replace them. Automate repetitive tasks to free up time for more meaningful and creative work.

8. Building a Resilient Organization

-**Adaptability:** Foster a culture of adaptability and resilience. Encourage employees to embrace change and continuously seek ways to improve and innovate.

-**Crisis Management:** Develop robust crisis management plans to ensure organizational stability during unexpected events. Provide support systems that help employees cope with crises.

-**Sustainability:** Implement sustainable practices that contribute to the long-term well-being of employees and the environment. Promote corporate social responsibility initiatives.

Chapter 9.
Ethics and Responsibility: AI in the Workplace

Ethical considerations and dilemmas

As the integration of automation and artificial intelligence (AI) continues to reshape the workplace, a host of ethical considerations and dilemmas emerge. These issues require careful attention to ensure that the implementation of new technologies aligns with societal values and promotes fairness, transparency, and human dignity. Here are some of the key ethical considerations and dilemmas associated with automation and AI:

1. Job Displacement and Economic Inequality

-Unemployment Risks: Automation can lead to significant job displacement, particularly in sectors reliant on routine tasks. This raises ethical questions about the responsibility of organizations and governments to support displaced workers through retraining programs and social safety nets.

-Widening Inequality: The benefits of automation often accrue to those who own and develop the technology, potentially widening economic inequality. Ensuring that the gains from automation are distributed more equitably is a major ethical challenge.

2. Bias and Fairness in AI Systems

-Algorithmic Bias: AI systems can perpetuate and even amplify existing biases if they are trained on biased data. This can lead to unfair treatment of certain groups, particularly in areas like hiring, lending, and law enforcement.

-Transparency and Accountability: Ensuring that AI systems are transparent and that their decision-making processes can be understood and challenged is crucial for maintaining fairness and accountability.

3. Privacy and Surveillance

-Data Privacy: The extensive data collection required for AI systems poses significant privacy risks. Organizations must navigate the ethical implications of data collection, ensuring that they respect individual privacy and comply with data protection regulations.

-Surveillance Concerns: The use of AI for monitoring and surveillance in the workplace can infringe on employees' privacy and create a climate of distrust. Balancing the benefits of surveillance with the need for privacy and autonomy is a key ethical dilemma.

4. Informed Consent and Autonomy

-Informed Consent: Ensuring that individuals are fully

informed about how their data will be used and obtaining their consent is essential. This includes clear communication about the implications of AI decisions on their lives and careers.

-**Employee Autonomy:** Automation can impact employee autonomy by dictating work processes and limiting creative input. Preserving a degree of human agency and decision-making is important for maintaining job satisfaction and ethical work practices.

5. Safety and Reliability

-**Ensuring Safety:** AI and automated systems must be designed to operate safely and reliably. This includes rigorous testing and validation to prevent accidents and harm, particularly in high-stakes environments like healthcare and autonomous vehicles.

-**Human Oversight:** Maintaining human oversight over AI decisions is crucial to catch errors and make ethical judgments that automated systems might overlook.

6. Ethical Use of AI in Decision-Making

-**Decision Transparency:** AI systems used in decision-making, such as hiring or criminal justice, must be transparent about their criteria and processes. Individuals affected by these decisions should have the opportunity to understand and contest them.

-**Ethical Programming:** Developers and organizations must ensure that AI systems are programmed with ethical considerations in mind, aligning their operations with societal values and norms.

7. Long-Term Implications and Responsibility

-**Future Workforce Planning:** Organizations have an ethical responsibility to consider the long-term implications of automation on the workforce. This includes planning for transitions and investing in employee development to prepare for future job roles.

-**Global Impact:** The ethical implications of automation extend beyond individual organizations and nations. Considering the global impact, including effects on developing economies and labor markets, is essential for responsible implementation.

8. Human-AI Collaboration

-**Complementary Roles:** Striking a balance between human and AI collaboration that maximizes strengths of both is crucial. Ensuring that AI supports human workers rather than fully replacing them can mitigate some ethical concerns.

-**Skill Development:** Providing opportunities for employees to develop new skills that complement AI is an ethical imperative to ensure that they remain relevant and valuable in the evolving job market.

9. Governance and Regulation

-Regulatory Frameworks: Developing robust regulatory frameworks that govern the ethical use of AI and automation is essential. These frameworks should address issues such as bias, privacy, and accountability.

-Industry Standards: Establishing industry standards and best practices for ethical AI use can help guide organizations in responsible technology implementation.

The ethical considerations and dilemmas associated with automation and AI are complex and multifaceted. Addressing these issues requires a collaborative effort between organizations, governments, and society to ensure that technological advancements promote fairness, transparency, and human dignity. By proactively engaging with these ethical challenges, we can harness the benefits of automation and AI while mitigating their potential harms, leading to a more equitable and just future.

Ethical considerations and dilemmas

Corporate Social Responsibility (CSR) refers to a business model in which companies integrate social and environmental concerns into their operations and interactions with stakeholders. It goes beyond profit-making to consider the broader impact of business activities on society and the environment. Here are the key elements and considerations of CSR:

1. Ethical Business Practices

-Integrity and Transparency: Companies should operate with integrity and transparency, ensuring honesty in all business dealings and communications. This builds trust with stakeholders, including customers, employees, investors, and the community.

-Fair Trade: Engaging in fair trade practices ensures that suppliers and producers, especially in developing countries, receive fair compensation. This contributes to global economic fairness and sustainability.

2. Environmental Responsibility

-Sustainable Practices: Implementing sustainable practices is crucial. This includes reducing carbon footprints, minimizing waste, and conserving natural resources. Companies can adopt renewable energy sources, improve energy efficiency, and engage in recycling programs.

-Eco-friendly Products: Developing and promoting products that are environmentally friendly can reduce the negative impact on the planet. This involves using sustainable materials, reducing packaging, and ensuring products are recyclable or biodegradable.

3. Community Engagement

-Philanthropy and Volunteering: Companies can support their communities through philanthropy, such as donations to local charities and nonprofits. Encouraging employee volunteering can also make a significant positive impact on the community.

-Local Economic Development: Investing in local communities by creating jobs, supporting local businesses, and enhancing infrastructure contributes to regional economic growth and stability.

4. Employee Welfare and Development

-Fair Labor Practices: Ensuring fair labor practices, such as providing safe working conditions, fair wages, and reasonable working hours, is fundamental to CSR. Protecting workers' rights and preventing exploitation are key ethical obligations.

-Professional Development: Offering opportunities for professional growth and development helps employees enhance their skills and advance their careers. Providing training, education, and career development programs benefits both employees and the company.

5. Consumer Protection and Satisfaction

-**Product Safety and Quality:** Ensuring that products and services are safe and of high quality is crucial. Companies must adhere to safety standards and regulations to protect consumers.

-**Honest Marketing:** Marketing practices should be honest and not misleading. Providing accurate information about products and services helps consumers make informed decisions and builds brand loyalty.

6. Governance and Accountability

-**Corporate Governance:** Strong corporate governance structures ensure that companies are accountable and operate in the best interest of their stakeholders. This includes having a diverse and independent board of directors and effective oversight mechanisms.

-**Stakeholder Engagement:** Actively engaging with stakeholders, including shareholders, employees, customers, suppliers, and the community, ensures that their interests and concerns are considered in decision-making processes.

7. Long-term Vision

-**Sustainable Growth:** Companies should focus on sustainable growth that benefits not only their bottom line but also society and the environment. This long-term vision involves balancing short-term gains with long-term impacts.

-Innovation for Good: Investing in innovation that addresses social and environmental challenges can lead to sustainable business models and new market opportunities.

8. Reporting and Transparency

-CSR Reporting: Regularly reporting on CSR activities and progress is important for transparency and accountability. This includes publishing sustainability reports that outline environmental, social, and governance (ESG) metrics and achievements.

-Third-Party Audits: Engaging third-party auditors to verify CSR claims and performance ensures credibility and trustworthiness. Independent assessments help identify areas for improvement and validate efforts.

Corporate Social Responsibility is a comprehensive approach that integrates ethical, environmental, and social considerations into a company's operations. By embracing CSR, companies can build trust with stakeholders, contribute to societal well-being, and ensure sustainable business practices. This commitment not only enhances the company's reputation but also creates a positive impact on the world, aligning business success with societal good.

Ensuring Fairness and Transparency

In the age of automation and artificial intelligence (AI), ensuring fairness and transparency in business operations and decision-making processes is more important than ever. These principles are critical for building trust, maintaining ethical standards, and promoting equity. Here are key strategies for organizations to ensure fairness and transparency:

1. Transparent Decision-Making

-**Clear Communication:** Communicate decisions and the rationale behind them clearly and openly with all stakeholders. Transparency in decision-making processes builds trust and reduces uncertainty.

-**Inclusive Processes:** Involve diverse stakeholders in decision-making processes to ensure that different perspectives are considered. This inclusivity helps to identify and mitigate potential biases.

2. Ethical Use of AI and Automation

-**Bias Mitigation:** Actively work to identify and eliminate biases in AI systems. This involves using diverse and representative datasets and continuously monitoring AI outcomes for fairness.

-**Explainability:** Develop AI systems that are explainable and understandable. Stakeholders should be able to comprehend how decisions are made and the factors that influence these decisions.

3. Fair Employment Practices

-**Equal Opportunities:** Ensure equal opportunities for all employees and applicants, regardless of their background. Implement policies and practices that promote diversity, equity, and inclusion.

-**Merit-Based Advancement:** Base hiring, promotion, and compensation decisions on merit and performance. Avoid favoritism and ensure that all employees have access to opportunities for growth and advancement.

4. Transparent Financial Practices

-**Accurate Reporting:** Maintain accurate and honest financial reporting. Regularly publish financial statements and reports that reflect the true state of the company's finances.

-**Open Access:** Provide stakeholders with easy access to financial information. Transparency in financial practices helps build investor confidence and supports informed decision-making.

5. Customer Transparency

-Honest Marketing: Ensure that marketing materials and advertisements are truthful and not misleading. Provide customers with accurate information about products and services.

-Clear Policies: Clearly outline policies related to returns, warranties, and customer service. Transparency in these areas helps build trust and improve customer satisfaction.

6. Data Privacy and Security

-Protecting Privacy: Implement robust data privacy practices to protect customer and employee information. Ensure compliance with data protection regulations such as GDPR or CCPA.

-Transparency in Data Use: Clearly communicate how data is collected, used, and shared. Obtain informed consent from individuals before using their data and provide options for them to control their data privacy settings.

7. Regulatory Compliance

-Adherence to Laws: Ensure compliance with all relevant laws and regulations. This includes labor laws, environmental regulations, and industry-specific standards.

-**Regular Audits:** Conduct regular internal and external audits to ensure compliance and identify areas for improvement. Transparency in regulatory compliance builds trust with regulators and the public.

8. Stakeholder Engagement

-**Open Dialogue:** Foster open dialogue with stakeholders, including employees, customers, investors, and the community. Regularly seek feedback and address concerns in a transparent manner.

-**Public Accountability:** Publicly disclose the outcomes of stakeholder engagement efforts and how feedback has been incorporated into decision-making processes.

9. Governance and Accountability

-**Strong Governance Structures:** Implement strong corporate governance structures to ensure accountability. This includes having a diverse and independent board of directors and effective oversight mechanisms.

-**Ethical Leadership:** Promote ethical leadership at all levels of the organization. Leaders should model transparency and fairness in their actions and decisions.

10. Social and Environmental Responsibility

-**Sustainability Reporting:** Regularly report on social and environmental performance, including sustainability initiatives and progress. Transparency in these areas demonstrates a commitment to corporate social responsibility.

-**Impact Assessments:** Conduct impact assessments to evaluate the effects of business activities on society and the environment. Use these assessments to inform strategies for minimizing negative impacts and enhancing positive contributions.

Ensuring fairness and transparency is essential for building trust and maintaining ethical standards in the modern business landscape. By adopting these strategies, organizations can foster a culture of integrity and accountability, promote equity, and enhance their reputation. Transparency and fairness not only contribute to long-term business success but also create a positive impact on society and the environment.

Chapter 10.
The Road Ahead: Preparing for an AI-Driven Future

Predictions and trends for the next decade

As we look towards the next decade, the rapid advancement of technology, particularly in the fields of automation and artificial intelligence (AI), will continue to drive significant changes across various industries and aspects of daily life. Here are some key predictions and trends that are expected to shape the future:

1. Accelerated Adoption of AI and Automation

-**Widespread Integration:** AI and automation technologies will become increasingly integrated into business operations, enhancing efficiency and productivity across sectors such as manufacturing, healthcare, finance, and retail.

-**AI-Driven Decision Making:** Businesses will rely more heavily on AI to analyze data and inform decision-making processes. This will enable more accurate predictions, personalized customer experiences, and optimized resource management.

2. Transformation of the Workforce

-**New Job Roles:** As automation takes over routine tasks,

new job roles that require advanced technical skills and creativity will emerge. There will be a growing demand for AI specialists, data scientists, cybersecurity experts, and professionals in emerging tech fields.

-**Reskilling and Upskilling:** Continuous learning and professional development will become essential. Organizations will invest in reskilling and upskilling programs to prepare employees for the evolving job market and ensure they remain competitive.

3. Growth of Remote and Flexible Work

-**Remote Work Norm:** Remote work, accelerated by the COVID-19 pandemic, will become a permanent feature for many organizations. Advances in collaboration tools and virtual environments will support seamless remote work experiences.

-**Hybrid Work Models:** Hybrid work models, combining remote and on-site work, will become more common. This flexibility will cater to employee preferences and enhance work-life balance while maintaining productivity.

4. Enhanced Healthcare Innovations

-**AI in Healthcare:** AI will revolutionize healthcare by enabling early diagnosis, personalized treatment plans, and predictive analytics. Technologies such as telemedicine, wearable devices, and AI-driven diagnostics will become standard.

-**Genomics and Precision Medicine:** Advances in genomics and AI will drive precision medicine, allowing for tailored treatments based on individual genetic profiles. This will lead to more effective and targeted healthcare solutions.

5. Evolution of Smart Cities

-**IoT and Connectivity:** The Internet of Things (IoT) will play a crucial role in developing smart cities, with interconnected devices enhancing urban infrastructure, transportation systems, and public services.

-**Sustainable Urbanization:** Smart cities will prioritize sustainability, incorporating green technologies, renewable energy sources, and efficient waste management systems to reduce environmental impact.

6. Financial Services Transformation

-**Fintech Growth:** Financial technology (fintech) innovations will continue to disrupt traditional banking. AI-driven financial services, blockchain, and digital currencies will reshape payment systems, lending, and investment management.

-**Enhanced Security:** As digital transactions increase, so will the emphasis on cybersecurity. Advanced security measures, including biometric authentication and AI-driven fraud detection, will become critical.

7. Advancements in Education

-**Personalized Learning:** AI will enable personalized learning experiences tailored to individual student needs and learning styles. Adaptive learning platforms and intelligent tutoring systems will enhance educational outcomes.

-**Lifelong Learning:** The concept of lifelong learning will gain prominence, with individuals continually updating their skills to stay relevant in a dynamic job market. Online courses, micro-credentials, and virtual classrooms will support this trend.

8. Sustainable Practices and Green Technologies

-**Environmental Responsibility:** Companies will increasingly adopt sustainable practices to address climate change and reduce their carbon footprint. Green technologies, such as renewable energy and electric vehicles, will gain widespread adoption.

-**Circular Economy:** The shift towards a circular economy, which emphasizes recycling, reusing, and reducing waste, will become more pronounced. Businesses will innovate to minimize environmental impact and create sustainable value chains.

9. Enhanced Consumer Experiences

-**Personalization and Customization:** AI will enable highly personalized and customized consumer experiences. From tailored product recommendations to bespoke services, businesses will leverage data to meet individual preferences.

-**Augmented and Virtual Reality:** Augmented reality (AR) and virtual reality (VR) technologies will transform how consumers interact with products and services. These immersive experiences will enhance shopping, entertainment, and training.

10. Ethical and Regulatory Considerations

-**Ethical AI:** The ethical implications of AI and automation will become a focal point. Companies and governments will work to develop frameworks that ensure ethical AI use, address biases, and protect individual rights.

-**Regulatory Landscape:** The regulatory landscape will evolve to keep pace with technological advancements. New regulations will emerge to govern data privacy, AI deployment, cybersecurity, and digital transactions.

The next decade promises significant technological advancements and transformative changes across various domains. By anticipating these trends and preparing for their impact, businesses, governments, and individuals can harness the potential of AI and automation to drive innovation, enhance quality of life, and create a more sustainable and equitable future.

Predictions and trends for the next decade

As automation and artificial intelligence (AI) reshape industries and redefine the nature of work, both individuals and organizations must adopt strategies to navigate and thrive in this evolving landscape. Here are key strategies for each:

Strategies for Individuals

1. Continuous Learning and Skill Development

-Embrace Lifelong Learning: Continuously seek opportunities to learn new skills and update existing ones. Online courses, certifications, and workshops can provide valuable knowledge and skills relevant to emerging technologies and trends.

-Focus on Technical Skills: Develop technical skills in high-demand areas such as AI, machine learning, data analysis, and cybersecurity. Proficiency in these fields can enhance employability and open new career opportunities.

-**Enhance Soft Skills:** Complement technical skills with strong soft skills such as critical thinking, problem-solving, communication, and teamwork. These skills are crucial for collaborating with AI systems and working effectively in diverse teams.

2. Adaptability and Flexibility

-**Stay Agile:** Be open to change and adaptable to new roles and responsibilities. Flexibility is key in a dynamic job market where job functions and requirements are constantly evolving.

-**Explore Multiple Career Paths:** Consider diversifying career paths and exploring roles in different industries. This can provide a broader range of opportunities and reduce reliance on a single career trajectory.

3. Networking and Collaboration

-**Build Professional Networks:** Actively network with professionals in your field and related industries. Networking can provide insights into emerging trends, job opportunities, and collaborative projects.

-**Engage in Collaborative Projects:** Participate in collaborative projects, hackathons, and industry events. These experiences can enhance skills, foster innovation, and build valuable connections.

4. Personal Branding and Visibility

-**Develop an Online Presence:** Create and maintain a professional online presence through platforms like LinkedIn. Showcase your skills, achievements, and projects to attract potential employers and collaborators.

-**Share Knowledge:** Contribute to industry discussions by writing articles, giving talks, or participating in webinars. Sharing knowledge can establish you as a thought leader and enhance your professional reputation.

Strategies for Organizations

1. Investing in Talent Development

-**Provide Training Programs:** Offer continuous training and development programs for employees to keep their skills up-to-date. This includes technical training, soft skills development, and leadership training.

-**Encourage Lifelong Learning:** Foster a culture of lifelong learning by supporting employees' pursuit of further education, certifications, and professional development opportunities.

2. Adopting Advanced Technologies

-**Implement AI and Automation:** Integrate AI and automation technologies to enhance efficiency, productivity, and innovation. Use these technologies to automate repetitive tasks and free up employees for more strategic work.

-**Leverage Data Analytics:** Utilize data analytics to gain insights into business operations, customer preferences, and market trends. Data-driven decision-making can improve outcomes and drive business growth.

3. Promoting an Inclusive and Flexible Workplace

-**Flexible Work Arrangements:** Offer flexible work arrangements, such as remote work and flexible hours, to attract and retain top talent. These arrangements can improve employee satisfaction and productivity.

-**Diverse and Inclusive Culture:** Promote diversity and inclusion in the workplace by implementing policies and practices that ensure equal opportunities and respect for all employees.

4. Enhancing Organizational Agility

-**Agile Methodologies:** Adopt agile methodologies to increase organizational agility and responsiveness to market changes. Agile practices encourage collaboration, iterative development, and continuous improvement.

-**Innovation and Experimentation:** Foster a culture of innovation by encouraging experimentation and risk-taking. Support employees in developing and testing new ideas and solutions.

5. Ethical and Responsible AI Use

-Ensure Ethical AI Practices: Develop and implement ethical guidelines for AI use to address issues such as bias, transparency, and accountability. Ensure that AI systems are fair, transparent, and aligned with ethical standards.

-Regulatory Compliance: Stay informed about regulatory developments related to AI and automation. Ensure compliance with relevant laws and standards to mitigate legal and reputational risks.

6. Strategic Planning and Future Readiness

-Future-Proofing Strategies: Develop strategic plans that consider long-term trends and potential disruptions. Identify future-proof skills and roles that will be essential as technology continues to evolve.

-Scenario Planning: Conduct scenario planning to anticipate various future scenarios and prepare for potential challenges and opportunities. This proactive approach can enhance organizational resilience.

For individuals, continuous learning, adaptability, networking, and personal branding are essential strategies to remain competitive in a rapidly changing job market.

For organizations, investing in talent development, adopting advanced technologies, promoting inclusivity and flexibility, enhancing agility, ensuring ethical AI use, and strategic planning are crucial for thriving in the era of automation and AI. By embracing these strategies, both individuals and organizations can navigate the future of work with confidence and success.

Building a Resilient and Adaptable Workforce

In an era characterized by rapid technological advancements and constant change, building a resilient and adaptable workforce is crucial for organizational success and sustainability. Here are key strategies to foster resilience and adaptability among employees:

1. Cultivate a Learning Culture

-**Encourage Lifelong Learning:** Promote a culture that values continuous learning and professional development. Provide access to training programs, online courses, workshops, and seminars that help employees acquire new skills and knowledge.

-**Support Skill Diversification:** Encourage employees to diversify their skill sets by exploring different areas of expertise. This can make the workforce more versatile and better prepared to handle various challenges.

2. Foster a Growth Mindset

-Promote a Growth-Oriented Attitude: Encourage employees to adopt a growth mindset, which emphasizes the importance of learning from failures, embracing challenges, and persisting in the face of setbacks.

-Recognize and Reward Effort: Acknowledge and reward not just the outcomes but also the efforts and improvements employees make. This can motivate them to continue developing their skills and taking on new challenges.

3. Provide Flexibility and Autonomy

-Flexible Work Arrangements: Offer flexible work options, such as remote work, flexible hours, and job-sharing arrangements. This flexibility can help employees manage their personal and professional lives more effectively, reducing stress and increasing job satisfaction.

-Empower Decision-Making: Give employees autonomy in their roles by empowering them to make decisions and take ownership of their work. This can enhance their sense of responsibility and confidence in handling diverse tasks.

4. Enhance Emotional Intelligence

-Emotional Intelligence Training: Provide training on emotional intelligence (EQ) to help employees improve their self-awareness, empathy, and interpersonal skills. High EQ can enhance teamwork and collaboration, especially during times of change.

-**Stress Management:** Offer resources and training on stress management techniques, such as mindfulness, meditation, and time management. Helping employees manage stress effectively can boost their resilience.

5. Encourage Collaboration and Teamwork

-**Collaborative Work Environment:** Foster a collaborative work environment where employees are encouraged to share ideas, support each other, and work together to solve problems. Teamwork can enhance creativity and innovation.

-**Cross-Functional Teams:** Form cross-functional teams to work on projects. This exposure to different perspectives and expertise can enhance employees' adaptability and problem-solving skills.

6. Invest in Leadership Development

-**Leadership Training Programs:** Provide leadership development programs to identify and nurture future leaders within the organization. Strong leadership is essential for guiding teams through change and uncertainty.

-**Mentorship and Coaching:** Establish mentorship and coaching programs where experienced leaders can provide guidance, support, and feedback to less experienced employees. This can help build a pipeline of resilient and adaptable leaders.

7. Foster a Positive Organizational Culture

-**Inclusive and Supportive Environment:** Create an inclusive and supportive workplace culture where employees feel valued and respected. An inclusive culture can enhance employee engagement and retention.

-**Transparent Communication:** Maintain open and transparent communication channels. Keeping employees informed about organizational changes, goals, and challenges can reduce uncertainty and build trust.

8. Implement Change Management Practices

-**Proactive Change Management:** Develop and implement effective change management practices to help employees navigate transitions smoothly. Provide training on how to handle change and uncertainty.

-**Feedback Mechanisms:** Establish feedback mechanisms to gather employee input on changes and initiatives. Involving employees in the change process can increase their buy-in and adaptability.

9. Promote Well-Being and Work-Life Balance

-Wellness Programs: Offer wellness programs that focus on physical, mental, and emotional well-being. Supportive health initiatives can improve overall employee resilience.

-Work-Life Balance: Encourage work-life balance by setting realistic expectations, promoting the use of vacation days, and discouraging excessive overtime. A healthy balance can prevent burnout and improve productivity.

10. Prepare for Future Trends

-Anticipate Industry Changes: Stay informed about industry trends and technological advancements that could impact the organization. Preparing for future changes can help employees adapt more quickly and effectively.

-Scenario Planning: Conduct scenario planning to identify potential future challenges and opportunities. Training employees to think strategically about the future can enhance their adaptability.

Building a resilient and adaptable workforce requires a multifaceted approach that includes promoting a learning culture, fostering a growth mindset, providing flexibility, enhancing emotional intelligence, encouraging collaboration, investing in leadership development, fostering a positive organizational culture, implementing change management practices, promoting well-being, and preparing for future trends. By adopting these strategies, organizations can create a workforce that is equipped to handle the uncertainties and opportunities of the future, ensuring long-term success and sustainability.

Chapter 11.
Conclusion: Embracing Change and Innovation

Recap of key insights

As we navigate the transformative era of automation and artificial intelligence (AI), several key insights emerge that are essential for understanding and preparing for the future of work. Here is a recap of these insights:

1. The Rise of AI and Automation

-**Integration Across Industries:** AI and automation are becoming integral across various sectors, enhancing efficiency, productivity, and innovation. Industries such as healthcare, finance, manufacturing, and agriculture are already seeing significant benefits from these technologies.

-**AI-Driven Decision Making:** AI systems are increasingly being used to analyze vast amounts of data, providing valuable insights that inform strategic decision-making and optimize operations.

2. The Evolving Workforce

-**New Job Roles:** The automation of routine tasks is leading to the creation of new job roles that require advanced technical skills and creativity. There is a growing demand for expertise in AI, data science, and other emerging fields.

-Reskilling and Upskilling: Continuous learning and skill development are crucial. Both individuals and organizations must invest in reskilling and upskilling to remain competitive in the evolving job market.

3. Work Flexibility and Remote Work

-Permanent Shift to Remote Work: The COVID-19 pandemic has accelerated the adoption of remote work, which is likely to remain a permanent feature for many organizations. Advances in technology will continue to support seamless remote work experiences.

-Hybrid Work Models: Organizations are increasingly adopting hybrid work models that combine remote and on-site work, offering greater flexibility and work-life balance for employees.

4. Innovations in Healthcare

-AI-Enhanced Healthcare: AI is transforming healthcare by enabling early diagnosis, personalized treatments, and predictive analytics. Telemedicine, wearable devices, and AI-driven diagnostics are becoming standard practice.

-Precision Medicine: Advances in genomics and AI are driving precision medicine, allowing for treatments tailored to individual genetic profiles, resulting in more effective healthcare solutions.

5. Smart Cities and Sustainability

-IoT and Smart Infrastructure: The Internet of Things (IoT) is playing a crucial role in developing smart cities, with interconnected devices enhancing urban infrastructure, transportation, and public services.

-**Environmental Responsibility:** There is a growing emphasis on sustainable practices and green technologies. Smart cities are prioritizing sustainability through renewable energy, efficient waste management, and reduced environmental impact.

6. Transformation in Financial Services

-**Fintech Disruption:** Financial technology (fintech) innovations are reshaping traditional banking. AI-driven financial services, blockchain, and digital currencies are revolutionizing payment systems, lending, and investment management.

-**Cybersecurity:** With the increase in digital transactions, the focus on cybersecurity is intensifying. Advanced security measures, including biometric authentication and AI-driven fraud detection, are becoming critical.

7. Education and Lifelong Learning

-**Personalized Learning:** AI is enabling personalized learning experiences tailored to individual needs and learning styles. Adaptive learning platforms and intelligent tutoring systems are enhancing educational outcomes.

-**Lifelong Learning:** The concept of lifelong learning is gaining prominence. Continuous education and professional development are essential for staying relevant in a dynamic job market.

8. Ethical and Regulatory Considerations

-**Ethical AI:** The ethical implications of AI and automation are becoming a focal point. Organizations must develop frameworks to ensure ethical AI use, address biases, and protect individual rights.

-**Regulatory Compliance:** The regulatory landscape is evolving to keep pace with technological advancements. New regulations are emerging to govern data privacy, AI deployment, and cybersecurity.

9. The Future of Work

-**Job Displacement and Creation:** Automation will displace certain jobs but also create new opportunities. Preparing the workforce for this transition through reskilling and upskilling is crucial.

-Gig Economy and Employment Models: The gig economy and new employment models are on the rise, offering more flexible work arrangements but also posing challenges related to job security and benefits.

10. Organizational Responsibility and Support

-Employee Well-Being: Organizations must prioritize employee well-being by providing resources and support systems to manage stress and maintain mental health.
Corporate Social Responsibility: Companies have a role in ensuring fairness, transparency, and ethical behavior. Corporate social responsibility initiatives can enhance reputation and contribute to societal well-being.

The future of work in the age of AI and automation presents both challenges and opportunities. By understanding these key insights, individuals and organizations can better prepare for and adapt to the rapidly changing landscape. Embracing continuous learning, fostering flexibility, prioritizing ethical practices, and investing in sustainable and inclusive strategies will be crucial for thriving in this new era.

The human element in the age of AI

As artificial intelligence (AI) and automation become increasingly integrated into our lives and workplaces, the importance of the human element cannot be overstated. While AI can handle a wide range of tasks, from data a nalysis to customer service, the unique qualities that humans bring to the table remain irreplaceable. Understanding and emphasizing these human elements is crucial for ensuring a harmonious and productive future in the age of AI.

1. Emotional Intelligence and Empathy

-**Human Connection:** AI lacks the ability to truly understand and respond to human emotions. Emotional intelligence (EQ) allows humans to build relationships, offer support, and create meaningful connections that machines cannot replicate.

-**Empathy in Healthcare:** In fields such as healthcare, empathy is essential. While AI can assist with diagnostics and treatment plans, the compassionate care provided by healthcare professionals is crucial for patient well-being and recovery.

2. Creativity and Innovation

-**Creative Problem-Solving:** Humans excel in creative thinking and problem-solving, often coming up with innovative solutions that AI might not generate. This creativity drives progress and leads to breakthroughs in various fields.

-**Art and Design:** In areas like art, design, and entertainment, human creativity and artistic expression are key. While AI can assist and inspire, the unique human touch and perspective are what make these creations special.

3. Ethical Judgment and Decision-Making

-**Complex Decisions:** Many decisions require ethical judgment and consideration of broader societal impacts. Humans can weigh moral implications, cultural contexts, and long-term consequences, something AI is not equipped to do effectively.

-**Ethical AI Deployment:** Ensuring that AI is developed and used ethically requires human oversight. Humans must establish guidelines and frameworks to address issues like bias, fairness, and accountability in AI systems.

4. Adaptability and Resilience

-**Navigating Uncertainty:** Humans are inherently adaptable and resilient, capable of navigating complex, uncertain, and rapidly changing environments. This flexibility is vital in industries undergoing constant transformation due to technological advancements.

-**Learning and Growth:** Human capacity for lifelong learning and personal growth allows individuals to continuously acquire new skills and adapt to new roles, ensuring relevance in a dynamic job market.

5. Interpersonal Skills and Collaboration

-Teamwork and Collaboration: Effective teamwork and collaboration are driven by strong interpersonal skills, including communication, negotiation, and conflict resolution. Humans excel in building cohesive teams and working together towards common goals.

-Leadership: Human leaders inspire, motivate, and guide teams, fostering a sense of purpose and direction. Leadership involves understanding human motivations and emotions, which are areas where AI falls short.

6. Cultural Sensitivity and Context

-Understanding Nuances: Humans can understand and appreciate cultural nuances, contexts, and diversity in ways that AI cannot. This sensitivity is crucial for global businesses and multicultural environments.

-Contextual Awareness: Humans are better at interpreting context and making decisions based on subtle cues and broader situational awareness. This ability is particularly important in fields like diplomacy, social work, and customer relations.

7. Ethical AI Development

-Inclusive Development: Humans must ensure that AI development is inclusive, considering diverse perspectives and needs. This involves actively preventing biases and ensuring that AI benefits all segments of society.

-Stakeholder Engagement: Engaging various stakeholders, including employees, customers, and communities, in discussions about AI implementation ensures that different viewpoints are considered, leading to more balanced and ethical outcomes.

The human element remains vital in the age of AI, bringing qualities that machines cannot replicate. Emotional intelligence, creativity, ethical judgment, adaptability, interpersonal skills, cultural sensitivity, and contextual awareness are areas where humans excel and will continue to be indispensable. As we move forward, the integration of AI should enhance and complement human capabilities, rather than replace them. Emphasizing and nurturing these human qualities will be key to building a future where technology serves humanity in meaningful and beneficial ways.

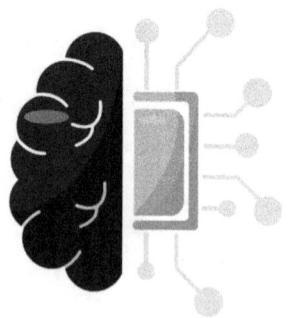

Final thoughts and calls to action

In conclusion, as we embrace the advancements of artificial intelligence (AI) and automation, it's crucial to recognize the indispensable role of the human element. While AI offers incredible potential to enhance efficiency, productivity, and innovation, it's the unique qualities that humans bring to the table that truly drive progress and ensure a harmonious future.

As we move forward, it's imperative that we prioritize the development and preservation of these human qualities. This includes nurturing emotional intelligence, fostering creativity and innovation, upholding ethical standards, cultivating adaptability and resilience, strengthening interpersonal skills, embracing cultural sensitivity, and championing inclusivity.

Moreover, it's essential that we take proactive steps to harness the power of AI responsibly and ethically. This involves establishing clear guidelines and regulations to govern AI development and deployment, ensuring transparency and accountability in AI systems, and actively addressing biases and inequalities.

As individuals, organizations, and societies, we have a collective responsibility to shape the future of AI in a way that prioritizes human well-being, fosters equitable opportunities, and promotes positive societal impact. By embracing the human element and working collaboratively towards these goals, we can build a future where AI serves as a tool for empowerment, innovation, and advancement.

Therefore, let us commit to nurturing our humanity amidst the technological revolution, advocating for ethical AI practices, and fostering a culture of inclusivity, empathy, and collaboration. Together, we can create a future where AI complements and enhances the human experience, leading to a more prosperous, equitable, and sustainable world for generations to come.

www.ingramcontent.com/pod-product-compliance
Lightning Source LLC
Chambersburg PA
CBHW070238230526
45470CB00002B/448